Henry Augustus Pilsbry, George Washington Tryon

Manual of Conchology, Structural and Systematic

Vol. 13

Henry Augustus Pilsbry, George Washington Tryon

Manual of Conchology, Structural and Systematic
Vol. 13

ISBN/EAN: 9783337803568

Printed in Europe, USA, Canada, Australia, Japan

Cover: Foto ©berggeist007 / pixelio.de

More available books at **www.hansebooks.com**

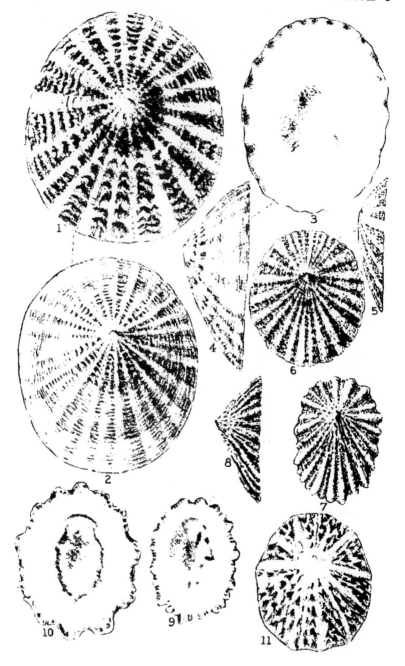

MANUAL

OF

CONCHOLOGY;

STRUCTURAL AND SYSTEMATIC.

WITH ILLUSTRATIONS OF THE SPECIES.

BY GEORGE W. TRYON, JR.

CONTINUED BY

HENRY A. PILSBRY.

Vol. XIII.

ACMÆIDÆ, LEPETIDÆ, PATELLIDÆ, TITISCANIIDÆ.

PHILADELPHIA:

Published by the Conchological Section,

ACADEMY OF NATURAL SCIENCES, COR. 19TH AND RACE STS.

1891.

In the present volume the important and difficult of group Doco-glossate Gastropods, the Limpets, is monographed, and in addition a small group not heretofore included. The material studied in the families *Acmœidœ* and *Patellidœ* is exceptionally extensive. A large number of forms are herein for the first time figured and adequately described. The value of the work has been enhanced by the liberality of Dr. W. H. Dall, of Washington, who placed at the author's disposal for study the magnificent collection of the Smithsonian Institution, a collection especially rich in species from the west coast of America, and containing the types of species described by GOULD, CARPENTER and DALL, many of which have not before been figured. No effort has been spared to make the synonymy and references complete and reliable; and it is hoped that conchologists will find the labor of classifying their collections of these intricate groups decidedly lightened.

Philadelphia, June, 1891.   H. A. P.

MANUAL OF CONCHOLOGY.

Monographs of the Acmæidæ, Lepetidæ, Patellidæ and Titiscaniidæ.

Family *ACMÆIDÆ* Cpr.

Acmæidæ CPR., Maz. Cat. p. 202, 1856.—*Tecturidæ* GRAY and authors.—*Lottiadæ* GRAY.—*Patellidæ*, in part, of authors.

Shell patelliform, conical, the apex more or less anterior, the embryonic shell conical, not spiral. Animal having a free branchial plume above the neck on the left side; radula without median teeth.

Animals of this family differ mainly from the *Patellidæ* and *Lepetidæ* in having a cervical branchial plume.

The shells may generally be known from *Patellidæ* by their different texture and the *more or less distinct internal border* of the aperture. They are never iridescent within.

They live on rocks and sea weeds, generally at very moderate depths. One species, *Acmæa fluviatilis*, is known to inhabit brackish water, and a few, like *Pectinodonta arcuata*, are abyssal.

The shells are excessively variable, as is usually the case in sedentary mollusks.

The author has examined very large suites of specimens, including nearly every species and variety described from the waters of North and South America, both east and west, and of Japan, Polynesia and Europe. The Australian and New Zealand forms are known to me by fewer specimens, and a number of the species of those regions I have not seen.

In the treatment of species I have aimed to be strictly conservative, reducing no described form to a variety or synonym without the most ample evidence of identity or intergradation of characters; and on the other hand, I have refrained from burdening science with new names for the vast number of transitional or divergent forms in the collections examined.

No characters diagnostic of the genera of Acmæidæ can be found in the shells.

(5)

Synopsis of genera.

I. *Radula with a single lateral tooth on each side; no uncini,*
PECTINODONTINÆ.

Genus PECTINODONTA Dall, 1882.

Animal blind; having a cervical branchial plume but no branchial cordon.

II. *Radula having three lateral teeth on each side,* ACMÆINÆ.

Genus ACMÆA Eschscholtz, 1830.

Animal having a cervical branchial plume but no branchial cordon; eyes present.

Genus SCURRIA Gray, 1847.

Animal having a cervical branchial plume and a complete or interrupted branchial cordon.

Subfamily PECTINODONTINÆ.

Genus PECTINODONTA Dall, 1882.

Pectinodonta DALL, Proc. U. S. Nat. Mus. 1881, p. 409, 1882; Blake Gastrop. p 411.

Shell resembling Scutellina but with a blunt subcentral apex. Soft parts resembling Acmæa except in the following details: Animal blind, with the front part of the head between the tentacles and above the muzzle much produced upward and forward, extending considerably farther forward than the end of the muzzle, which is marginated with lappets at the outer corners. Jaw thin, translucent. Gill exactly as in Acmæa; sides of foot and mantle edge simple, nearly smooth. Dental formula 0 (1.0.1.) 0; teeth large, with transverse pectinated or denticulated cusps, the serrated edge of which is turned toward the median line. The number of teeth is the smallest in any known limpet. (*Dall.*)

The dentition is figured on pl. 33, fig. 74.

P. ARCUATA Dall. Pl. 33, figs. 74, 75, 76.

Shell white, elongate-ovate, moderately elevated, with a blunt polished apex, on which in young specimens, remain traces of the disk-like, chalky, embryonic shell; the slopes from the apex to the ends both convexly arched; margin simple or slightly denticulated by the radiating sculpture; within polished; scars as in Acmæa;

epidermis none; sculpture externally of fine, uniform, rounded, closely set threads, radiating from near the apex to the margin and reticulated by the fine, rather prominent, regular, concentric ridges of growth, both ridges and threads averaging near the margin about three and a half to the millimeter. Length., from end to end, 14·5 mm.; from apex to anterior end 5·5 mm.; lat. 10·0 mm.; alt. 5·5 mm. (*Dall.*)

Off St. Lucia, 226 fms.; *off Dominica*, 333 fms.; *off Guadelupe*, 583 fms.; *and off St. Thomas.*

P. arcuata DALL, Proc. U. S. Nat. Mus. 1881, p. 409, 1882; Blake Rep. p. 411, t. 25, f. 3, 3a, 3b.

Subfamily ACMÆINÆ.

Genus ACMÆA Eschscholtz, 1830.

Acmæa ESCH., in append. Kotzebue's Neue Reise, ii, p. 24, 1830, type *A. mitra* Esch.—FORBES & HANLEY, Brit. Moll., ii, p. 433.— CPR., Mazat. Catal. p. 202.—DALL, Amer. Journ. Conch. vi, p. 237.—WATSON, Challenger Gastr. p. 28.—FISCHER, Manuel, p. 865.—*Tectura* AUD. & MILNE-EDW., in Cuvier's Rapport sur trois Mémoires, etc., Annales des Sci. Nat. xxi, 1830, p. 326, published not before 1831, type *P. virginea.*—*Tectura* of GRAY, H. & A. ADAMS, JEFFREYS, *et al.*—*Patelloidea* QUOY & GAIMARD, Voy. Astrol. iii, p. 349. Type *P. fragilis* (Chemn.) Q. & G., 1834.— *Lottia* GRAY, in parc, Philos. Trans. 1833, p. 800.—*Lottia* of GOULD, *et al.*—*Erginus* JEFFREYS Ann. Mag. N. H. 4th ser., xix, p. 231, March, 1877. Type *Tectura rubella* Fabr.—*Collisella* DALL, Amer. Journ. Conch. vi, p. 245, 1871. Type *A. pelta* Esch.—*Collisellina* DALL, *l. c.*, p. 154, type *A. saccharina.* L.

Shell conical, patelliform, apex more or less anterior. Animal with a branchial plume at the left side of the neck above; no branchial cordon. Dentition, see below.

The shells may generally be distinguished from Patella by the different texture and marginal border of the inside.

The thorough discussion of the generic name of this group contained in the various papers of Dr. Dall and others, renders any justification of the view of its nomenclature here taken, unnecessary.

Tectura and *Erginus* must be regarded as absolute synonyms of *Acmæa s. str.; Patelloidea* Q. & G. will probably be found to differ somewhat anatomically, and may then be utilized for a subgeneric group. The subgenera *Collisella* and *Collisellina* are defined below.

Species of the genus Acmæa are found in the littoral and laminarian zones of nearly all seas, except the waters adjacent to the continent of Africa.

The shells are subject to even greater mutations than the Patellidæ, and species are correspondingly difficult to define and limit. More than any other shells, these must be studied with constant reference to not only habitat geographically, but *station* as well. For an exact knowledge of the group we must therefore wait until observations on the species are made with especial reference to their modes of life and surroundings. Such data should be attached to every limpet collected.

Acmæa has been divided by Dr. W. H. Dall into a number of groups which may be tabulated as follows:

A. Muzzle with lappets; no uncini; formula of teeth 0 (3.0.3) 0,

$$\textit{Acmœa s.s.}$$

B. Muzzle without lappets, uncini present, *Collisella* Dall.

    a. formula of teeth 1 (3.0.3) 1   *Collisella* s.s.

    b. formula of teeth 2 (3.0.3) 2   *Collisellina* Dall.

The type of *Acmœa* is *A. mitra* Esch., dentition pl. 42, fig. 82; of *Collisella, A. pelta* Esch., dentition pl. 42, fig. 81; and the type of *Collisellina* is *A. saccharina* L., dentition pl. 42, fig. 83.

It is practically impossible at present to group the species of the entire world according to anatomical characters, or to decide to what extent these divisions will prove applicable to the entire series.

The most convenient and in most cases the most natural division of the genus is geographic. Thus considered, the species fall into six groups:

    I.   North Atlantic and Arctic.

    II.   Western coast of North America.

    III.   Western coast of South America.

    IV.   West Indies.

    V.   Japan.

    VI.   Indo-Pacific.

(VII. Species of unknown habitat.)

Of these groups, the second has great affinity to the first and fifth; the fourth may be regarded as derived from the second during the early tertiary period.

I.   Species of European Seas and the North Atlantic.

Many specimens of all of the species of this region have been examined by me.

A. RUBELLA Fabricius. Pl. 42, figs. 79, 80.

Shell small, rounded-oval, conical, apex elevated, situated at the anterior fourth of the shell's length; front slope steep, straight or concave, posterior slope convex. Surface smooth, showing faint lines of growth. Color reddish-buff or orange; inside of the same color, the border flesh-colored. Length 5, breadth 4, alt. 2½ mill.

Finmark, Norway; Greenland, 5–40 fms.

Patella rubella FABR., Fauna Grönl., p. 386, 1780.—*Tectura* (*Erginus*) *rubella* JEFFREYS, Ann. Mag. N. H., Mar., 1877, xxi, p. 231.—SARS, Moll. Arct. Norv. p. 121, t. 8, f. 5; t. ii, f. 11 (dentition), 1878.—*Pilidium fulvum,* in part, DALL, Am. Journ. Conch. v, 1869.—*Acmæa rubella* DALL, Proc. U. S. Nat. Mus. 1879, p. 337.

This little shell is of a more erectly conical form than *Pilidium fulvum,* and lacks radiating sculpture. It is smaller than *A. virginea* and not radiately painted, besides having the summit more anterior and more elevated. The specimens before me are from Greenland. It has been reported from the New England coast, but I am not sure of the correctness of the determination.

A. VIRGINEA Müller. Pl. 10, fig. 13, 14.

Shell small, oval, conical; apex at or a little back of the anterior fifth of the shell's length. Surface having delicate, almost obsolete radiating striæ and delicate growth lines. Color a delicate pink, with numerous (about 13) pink rays. Upper part of the cone buffish-white.

Inside pink or white, center flesh-colored or opaque white.

Length 9–10, breadth 6¾–9, alt. 3½–4 mill.

Mediterranean and Adriatic Seas; Atlantic from Norway and Iceland to the Canaries, Azores and Cape Verde Is., low water to 60 fms.

Patella virginea MULL., Zool. Dan. Prodr. i, p. 43, 1776.—GMEL., Syst. Nat. xiii, p. 3711.—*Acmæa virginea* HANLEY, Br. Mar. Conch. p. 32, 1844.—FORBES & HANLEY, Hist. Brit. Moll. ii, p. 437, t. 61, f. 1, 2.—DALL, Am. Journ. Conch. vi, p. 243, 1871.—BUQUOY, DAUTZ, & DOLLF., Moll. du Rouss. p. 478, t. 51, f. 12, 13.—*Tectura virginea* JEFFR., Brit. Conch. iii, p. 248; v, p. 200, t. 58, f. 4.— SARS, Moll. Reg. Arct. Norv. p. 121, t. ii, f. 10 (dentition).—*Patella parva* DA COSTA, Brit. Conch. p. 7. t. 8, f. 11, 1778, of DONOVAN and MONTAGU.—*Lottia unicolor* FORBES, Rep. Æg. Invert. pp. 135, 188, 1844.—*L. pulchella* FORBES, *l. c.,* p. 137.—*Lottia pellucida*

WEINKAUFF (not Linné) Jdurn. de Conchyl. x, p. 334, 1862.—
Patelloidea virginea COLBEAU, Moll. viv. de la Belg., p. 14.—*Patella
æqualis* Sow., Min. Conch. t. 139.—*Patella astensis* BONELLI.

A small and delicate species, pink rayed on a pale ground. It is
widely distributed in European seas. The following mutations
have received names:

Form *conica* Jeffr. Smaller than the type, more conical, summit
more elevated, nearly central. This form is figured by Wood,
Crag Moll. pl. 18, f. 6ᴄ.

Form *rotundata* Monts. More rounded than the type.

Form *depressa* Wood. Crag Moll. pl. 18, f. 6ᴀ.

Form *unicolor* Forbes. Of a uniform rosy color, without rays;
small.

Form *lactea* Jeffr. Milky-white.

A. TESTUDINALIS Müller. Pl. 9, figs. 25, 26, 27, 28, 29.

Shell conical, oval, the apex a little in front of the middle; posterior slope slightly convex, other slopes straight; surface more or
less distinctly, finely radiately striated; color yellowish-gray, with
numerous blackish-brown stripes, generally broken into a coarse
network, or tessellated pattern.

Inside white, with a large dark brown central area, the border
tessellated brown and white. Length 38, breadth 28, alt. 13 mill.

*North Atlantic and Arctic Oceans, southeast to the English Channel,
southwest to Long Island Sound; North Pacific from Sitka (and
Yesso?) to the Arctic Ocean.*

Patella testudinalis MULL., Prodr. Zool. Dan. p. 237, 1766.—
REEVE, Conch. Icon. f. 70.—*Tectura testudinalis* JEFFREYS, Brit.
Conch. iii, p. 246; v, p. 200, t. 58, f. 3.—GLD., Invert. of Mass.,
Binney's edit., p. 267, f. 529.—*Acmæa testudinalis* FORBES & HAN-
LEY, Hist. Brit. Sh. ii, p. 434, t. 62, f. 8, 9; t. AA, f. 2 (animal).—
DALL., Amer. Journ. Conch. vi, p. 249, t. 14, f. 13 (dentition);
Proc. U. S. Nat. Mus. 1878, p. 339.—SARS, Moll. Reg. Arct. Norv.
p. 120, t. ii. f. 9 (dentition).—*Lottia testudinalis* FORBES, Malac.
Monensis p. 34.—GLD., Invert. of Mass., 1st edit., p. 153, f. 12.—
Patella testudinaria and *P. tessellata* MULL.—*P. clealandi* SOWB.,
Trans. Linn. Soc. xi, p. 621.—*P. amœna* SAY, Journ. Acad. N. S.
Phila. ii, p. 223.—DE KAY, N. Y. Moll. p. 162, t. 9, f. 196.—*P. cly-
peus* BROWN, Ill. Conch. Gt. Br. t. 37, f. 9, 10.—*Patella alveus* CON-
RAD, Journ. Acad. N. S. Phila. vi, p. 267, t. 11, f. 20, 1831.—*Patel-*

loidea alveus COUTH., Bost. Journ. N. II, ii, p. 177.—*Lottia alveus* GLD., Inv. of Mass., p. 154, f. 13.—*Tectura alveus* BINNEY in GOULD, Inv. of Mass., 2d edit., p. 269, f. 530.—*Acmæa testudinalis* var. *alveus* DALL, Proc. U. S. Nat. Mus. 1878, p. 340.

Atlantic specimens of this well-known shell, although very variable in size and coloration, are readily distinguished from the other forms, the only considerable divergence being found in var. *alveus*. Specimens from the Aleutian Is., according to Dr. Dall, completely bridge the gap between *testudinalis* and *patina*. I have retained the latter separate, simply as a matter of convenience; but I do not doubt that it must be regarded as a geographic race of *testudinalis*.

American specimens are generally larger than European; figs. 27-29 represent specimens from Maine, figs. 25, 26 from England.

Var. ALVEUS Conrad. Pl. 42, figs. 90, 91.

Small, thin, compressed at the sides; apex acute and a little hooked forward. Surface delicately striated, interruptedly striped or tessellated with brown. Inside showing the markings of the exterior.

Massachusetts to Arctic Ocean; Sitka northward.

Numerous transitions occur between this and typical testudinalis. The narrow form is caused by the residence of individuals on seaweed or *Zostera* fronds.

* *_{*} *

II. SPECIES OF THE WESTERN COAST OF NORTH AMERICA.

The author has examined specimens of all of the species of this region, including many original types. Of most species many hundreds of shells have been studied. The elaborate papers of Dr. P. P. Carpenter, and of Dr. Wm. H. Dall have been freely used.

A. PATINA Eschscholtz. Pl. 2, figs. 34, 35, 36, 37 ; pl. 9. figs. 6–14.

Shell large, oval or rounded-oval, depressed-conic, the apex rounded and near the middle; slopes slightly convex. Surface obsoletely radiately striated, olive-gray, tessellated, or more rarely striped, with black.

Inside white with an irregular brown central area and a rather wide dark or tessellated border. Length 53, breadth 46, alt. 18 mill.

Aleutian Is. to San Diego, California.

A. patina Esch., Zool. Atlas, edit. Rathke., p. 19, t. 24, f. 7, 8.—
Midd., Sib. Reise, p. 187, t. 16, f. 1a–d, 2a–c, 3.—Cpr., Mazat. Cat.
p. 207 ; Amer. Journ. Conch. ii, p. 333.—*A. scutum* Esch., not
Orb.—*P. mammillata* Nutt., Jay's Catal. no. 2839.—Rve., Conch.
Icon f. 140.—*P. tessellata* Nutt., Jay's Cat. no. 2885.—*P. fenestrata*
Nutt., Rve. Conch. Icon. f. 121.—*P. verriculata* Rve., *l. c.*, f. 87.—
P. nuttalliana Rve., *l. c.*, f. 81.—*P. cumingii* Rve., *l. c.*, f. 37.—*Lot-
tia pintadina* Gould, U. S. Expl. Exped. t. 29, f. 455.—*Collisella
patina* Dall, Amer. Journ. Conch. vi, p. 247, t. 14, f. 4 (dentition).—
A. testudinalis var. patina Dall, Proc. U. S. Nat. Mus. i, p. 340.

P. cinis Rv., considered a synonym of *patina* by Cpr., belongs to
A. pelta. *P. strigillata* Nutt. mss. is a form of *fascicularis*, judging
from the suite deposited by Nuttall in the Academy collection.

This is the commonest of all western limpets. Although it has
been shown to intergrade with *A. testudinalis* on the Alaskan coast,
yet I cannot rank it as a variety of that species in the sense in
which *alveus* is a variety. It is thoroughly differentiated from *testudinalis* throughout most of its range. The two forms vary in quite
diverse directions, *patina* having no form corresponding to the var.
alveus of *testudinalis*, but having its own peculiar mutations, not
found in the other species.

It would be an advantage if we were to use the term "form"
(*forma*) for such mutations as *alveus*, *nacelloides*, etc., reserving the
rank of "variety" for true geographic subspecies.

The principle mutations of *A. patina* are as follows:

Var. PINTADINA Gld. (pl. 9, fig. 6). Large, flat, open, apex
subcentral ; tessellated white and dark. *P. cumingii* Rv. (pl. 42,
fig. 87) and *tessellata* Nutt. belong here as synonyms. This form
passes into the striped form *nuttalliana* Rve. (pl. 2, figs. 32, 33, and
also f. 36, 37). The last figures correspond to Reeve's *verriculata*.

Another mutation is the form *fenestrata* Nutt. (pl. 9, figs. 10, 11,
12, 13, 14), of which *cribraria* Gld. mss. is a synonym. This shell
when young is dark olive closely dotted all over with white, the
eroded apex black ; when adult it is usually uniform dull slate-color
outside with a ring of light around the black apical spot ; inside it
has a wide dark border, a large, irregular central dark patch, and
generally is suffused with dark brown all over. Sculpture obsolete.
This form is from San Francisco, Santa Cruz, etc.

Var. OCHRACEA Dall (pl. 9, figs. 7, 8, 9). Externally of a very
light yellowish-brown, without spots or rays ; internally white with

the characteristic dark brown stain of *patina* in the visceral area. The exterior is covered with fine, regularly radiating, close, equal, thread-like riblets, which pass from apex to margin without bifurcation, imbrication or asperities of any kind. These riblets will serve to distinguish it from any of the other limpets of the coast; otherwise it approaches very close to some varieties of *scabra* and can be traced right into some varieties of *patina*. (*Dall.*)

This variety was described from Monterey, Cal.; it has also been found on Vancouver Id.

A. DALLIANA Pilsbry. Pl. 7, figs. 57, 58, 59, 60.

Shell large, oblong, depressed, rather thin. Apex low, curving forward; length of front slope contained about 3½ times in the length of the shell; posterior slope gently convex. Surface covered with close, slightly unequal radiating riblets, each rendered rasp-like by very close, regular and erect delicate lamellæ; interstices narrow, having growth-striæ but no lamellæ.

The color is chestnut-brown, becoming dark umber in places, having short streaks and spots of white, forming a sparse tessellation. Inside light blue, with a small brown spot at the cavity of apex, and showing the color-pattern of the outside faintly through. Border wide, deep brown with white spots.

Length 46, breadth 32, alt. 6½ mill.

Angel Island, Porto Refugio, Gulf of California.

This is one of the finest American *Acmæas.* The oblong, somewhat parallel-sided and depressed contour, thin texture, and the beautifully sharp and regular file-like sculpture of the low, close riblets, are its prominent features. It is allied to *A. scabra*, but the enormous number of specimens of that species which I have examined in the Philadelphia and Washington collections, furnish no forms leading toward the *Dalliana.* The species is named in honor of Dr. Wm. H. Dall, who outlined the classification of the Acmæidæ in essentially its modern form, twenty years ago.

A. SCABRA Reeve. Pl. 3, figs. 38–49.

Shell thin, rounded-oval, depressed; apex situated between the center and the anterior third; surface sculptured with close, fine, minutely scaly riblets, of which larger ones are placed at regular intervals. Color light yellow, indistinctly spotted (rarely striped in divaricating pattern) with brown.

Interior porcelain-white or blue-tinted, with sometimes a few faint spots of brown in the cavity. Inside border transparent-yellowish or showing faint brown markings.

Length 37, breadth 31, alt. 7–8 mill.

Vancouver's Island to Acapulco, western Mexico.

Patella scabra Rv., Conch. Icon. f. 119.—*Acmæa scabra* CPR., Am. Journ. Conch. ii, p. 340.—*Collisella scabra* DALL, *l. c.*, vi, p. 251, t. 14, f. 12, 12a (dentition).—*Acmæa (scabra var.?) mörchii* DALL, Proc. U. S. Nat. Mus. 1878, p. 47.

The typical form is easy to recognize by its light coloration and the fine rasp-like riblets of the surface. Forms in which the outer layer of the shell is deep brown instead of buff, and the inside border consequently blackish, are quite similar to some variations of *A. patina ;* but sculpture and color-pattern will usually permit one to separate them readily.

Two forms have been described :

Var. LIMATULA Cpr. Pl. 3, figs. 38, 39, 40.

Outer layer of the shell black, covered with an olive-green, or sometimes light bluish, epidermis ; inside border black ; a deep brown central spot. Distribution mainly southern, San Diego to Acapulco.

A very beautiful color-pattern is shown in figs 45, 46, drawn from San Diego specimens. White rays alternate with dark olive.

As an extreme form of this variety, Var. MORCHII of Dall (pl. 3, figs. 47, 48, 49), must be ranked. It is typically much elevated, the apex subcentral and curved forward, sculpture coarse. Otherwise like var. *limatula.* Locality, Tomales Bay, Lower California. The large suite of shells before me from Tomales Bay show every intermediate stage between the high, cap-shaped forms and the normal *limatula.* The former constitute a peculiar phase of development attained by comparatively few individuals. Figures 47–49 are drawn from Tomales Bay specimens.

A. SPECTRUM Reeve. Pl. 1, figs. 7, 8, 9.

Apex rather anterior ; slopes rather straight ; sculptured with very strong close rough ribs, with smaller intervening riblets ; center of the inside white, with dark spots and bars.

Normally it is solid, rather depressed, with from 20–30 very strong, rounded ribs not evanescent anteriorly, the interstices being occupied by intercalary riblets. The color is white, with fine lines of brown (not striped as in *pelta* and *persona*) between the principal ribs, which delicately dot the otherwise uniform white margin. Sometimes the principal ribs are rather sharp, palmating the margin, occasionally they are small and crowded, becoming faint at the margin, when the shell presents the internal aspect of *A. mitella;* at other times assuming that of *Patella pediculus.* Generally the apex is at the anterior third; rarely at the anterior fourth, with very elongated outline; but sometimes is nearly central, with a rounded shell. In this species also there is occasionally found a var. *textilis;* when the ribs become faint and distant, the color-lines run into network, and the shell is of a thinner texture. The young is extremely inequilateral, and rapidly developes the characteristic ribs. Inside the shell has a white callus, through which the dark irregular blotch appears. This occasionally takes the form of irregular ghostly bars, which gave the name to the species. (*Cpr.*)

Length 34, breadth 24, alt. 12 mill.

Bodega Bay and San Francisco south to Lower California.

Patella spectrum Rv., Conch. Icon. f. 76.—*Acmœa spectrum* Cpr., Amer. Journ. Conch. ii, p. 339.—*Collisella spectrum* Dall, Amer. Journ. Conch. vi, p. 251, t. 14, f. 10 (dentition).—*Lottia scabra* Gld. (part), Expl. Exped. Shells.

The very strong ribs of the outside, and the curiously marked interior, like print of a hand, are prominent characters of this species.

It belongs to a group of forms represented in South America by *A. variabilis* and *A. ceciliana;* in China by *A. hieroglyphica*, and in Australasia by *A. marmorata*, etc. All showing curiously figured interiors.

A. persona Eschscholtz. Pl. 2, figs. 25, 26, 27, 28; pl. 3, figs. 51–56.

Shell oval, apex pointing forward, posterior slope long, convex, anterior slope short. Sculptured with strong, rounded ribs, usually nodulous, but sometimes obsolete. Whitish, with stripes and zigzags of blackish-brown, or olive-green variegated and speckled with white. Margin crenated by the ribs.

Inside white or stained with yellowish-brown, with a large central deep brown area, rarely absent; border articulated black and gray

Sitka to Turtle Bay, L. California.

A. persona Esch., Zool. Atl. v, p. 20. no. 9, t. 24, f. 1, 2.—Cpr., Amer. Journ. Conch. ii, p. 337.—*A. ancylus* Esch., *l. c.*, t. 24, f.

4–6.—*A. digitalis* Esch., *l. c.*, t. 23, f. 7, 8.—*P. umbonata* Nuttall, in Rve., Conch. Icon. f. 107.—*P. oregona* Nutt., *l. c.*, f. 112.—*L. textilis* Gould, Expl. Exped. Sh. t. 29. f. 456.— *Collisella persona* Dall, Amer. Journ. Conch. vi, p. 250, t. 14, f. 8 (dentition).— *L. scabra* Gld. (in part), Expl. Exped. Sh., f. 456b.—*A. radiata* Esch., Zool. Atl. p. 20.—*Tectura persona* Martens, Mal. Bl. xix, p. 95, t. 3, f. 5, 6.

An excessively variable species, ranging from about 30° to 50° N. lat.

The typical PERSONA is rather a smooth shell, corresponding to figs. 51, 52, of plate 3.

Two main races may be distinguished. The minor modifications of each are numberless.

Var. DIGITALIS Esch. Pl. 2, figs. 29, 30, 31 ; pl. 3, figs. 53, 54, 55, 56.

This is the most usual form found north of San Francisco Bay. It is dull, lusterless, whitish, with stripes and zigzags of blackish-brown. The apex is usually decidedly anterior and elevated ; the front ribs are obsolete, the posterior ribs strong, rounded, often uneven. Inside margin conspicuously tessellated ; central area generally dark and rather narrow. This is the *oregona* of authors, and probably *radiata* of Eschscholtz. It resembles the striped variety of the Chilian *A. ceciliana* so closely that it would be absolutely impossible to separate a mixed lot.

Var. UMBONATA Nuttall. Pl. 2, figs. 25, 26, 27, 28.

The prevalent form southward of San Francisco is an oval shell with rather spreading sides, the ribs narrow, interspaces wide and flat. Color dark olive to blackish, closely flecked with fine white dots, and usually having coarse white dashes also.

This variety becomes at times wholly free from ribs.

Another variety, typically equally distinct, but nameless, is found rom San Francisco to San Diego. It is a small shell resembling somewhat *A. patina*. There are no riblets. The surface is lusterless, white, with numerous, rather narrow, radiating brown stripes, often broken or abruptly divaricating. Inside generally without a central dark area. Gould's figures of the synonymous *L. scabra* are copied on pl, 29, figs. 47, 48, 49.

A. PELTA Eschscholtz.   Pl. 8, figs. 86–95.

Shell oval, conical, apex a little in front of the middle. Surface having rather coarse low ribs. *Dark border of the inside very narrow, or reduced to a series of dark scallops.*

Aleutian Is. and south coast of Alaska to the Santa Barbara Islands, California.

A. pelta Esch., Zool. Atl. pt. v, p. 19.—CARPENTER, Amer. Journ. Conch. ii, p. 336.—DALL, Proc. U. S. Nat. Mus. 1878, p. 338.—*Patella fimbriata* GLD., U. S. Expl. Exped. atlas, f. 445.— *P. leucophæa* (Nutt.) RVE., Conch. Icon. f. 101.—*P. monticola* NUTT., mss.—? *A. cassis* ESCH., Zool. Atl., p. 19, t. 24, f. 3.—? *A. pileolus* MIDD., Beitr. zu Mal. Ross. ii, p. 38, t. 1, f. 4, *teste* Cpr.—*Collisella pelta* DALL, Amer. Journ. Conch. vi, p. 246, t. 14, f. 6 (dentition).— *Tectura cassis* MARTENS, Mal. Bl. xix, p. 92, t. 3, f. 9, 10.—*Patella cinis* RVE., Conch. Icon. f. 60a, b, c.—*A. pelta var. nacelloides* DALL., Amer. Journ. Conch. vi, p, 247, t. 17, f. 36.

Prominent characters of this species are the erect, conical form, rather wide coarse ribs, and the narrow margin of the inside, usually not continuous but composed of scallops or square spots.

The variations may be classed under two main groups, as follows. Numerous intermediate forms occur.

(1) Var. PELTA Esch., typical.   Pl. 8, figs. 90, 91.

Rather large, solid, strong, with low coarse ribs, almost obsolete, or visible only posteriorly. Central dark spot of the interior rather small or wanting. Grayish-white, with numerous radiating black stripes, often divaricating or broken into a tessellated pattern.

As the ribs become stronger this passes into—

Form *cassis* (Esch.) Martens.   Pl. 8, figs. 86, 87, 88, 89.

Solid, strong, having stout radiating ribs about 25–27 in number, those in front narrower or obsolete. Dark spot of the inside small or obscured ; margin with a mere dark line, or a series of scallops between the ends of the ribs. Outside dull, grayish.

Another form connecting with the typical *pelta* is figured on pl. 8, figs. 92, 93, 94. It is small, conical, elevated, having much the shape of *A. mitra*. The color outside is gray, pink or light purple, painted with few or many black stripes. A dark spot is inside. Ribs obsolete. This is common at Olympia, Washington, growing on *Mytilus*. See Hemphill, Proc. A. N. S. Phila., 1881, p. 88.

2

(2) Var. NACELLOIDES Dall (Pl. 6, figs. 43, 44, 45), agrees with *A. instabilis* in the blackish-brown color and in sculpture, but it is less compressed laterally, and the basal margins are level, not elevated at the ends. It is abundant, living on kelp, at Monterey.

The proof of the alleged specific identity of *instabilis* with *pelta* is incomplete. The specimens collected by Henry Hemphill and described by him in Proc. Acad. Nat. Sci. Phila. 1881, p. 87, under the name *instabilis*, are typical *nacelloides*. Hemphill found that when these limpets live on the fronds and stems of kelp (*Phyllospora*) they have always the *Nacella*-like form and are black or dark brown; when an individual leaves the kelp for a station on the rocks its additional growth is of the normal black and white striped or tessellated pattern usual in typical *A. pelta.* A specimen of this form, beginning life as *nacelloides* and becoming *pelta*, is figured on pl. 8, fig. 95.

A. INSTABILIS Gould. Pl. 6, figs. 32, 33.

Shell narrow and oblong, the basal margin elevated at the ends; texture thin; slopes convex or bulging. Surface finely radiately striated; dark brown or black. Inside white or bluish, with or without a faint brown spot in the cavity.

Large specimens measure 1½ inches (38 mill.) in length by ⅞ in. (23 mill.) breadth; but the usual length is about one inch.

Vancouver Id. to Monterey, Cal.

P. instabilis GLD., Proc. Bost. Soc. N. H. ii, p. 150, 1846; U. S. Expl. Exped. Atlas f. 454, 454a.—*Nacella instabilis* CPR., et al.— *Acmœa instabilis* DALL, Amer. Journ. Conch. vi, p. 245.

This species has been considered by some to be a form of *A. pelta*; but no specimens connecting the two species have been reported, although it is not at all improbable that such may occur. At present *A. instabilis* has as valid grounds for being retained as a distinct species as *A. insessa* or *A. asmi.*

A. INSESSA Hinds. Pl. 6, figs. 36, 37.

Shell rather thin but strong; outline oval or oblong, the sides often parallel; elevated, conical, the apex in front of the middle, slopes convex; surface smooth, polished. Color varying from yellowish or olive-brown to chocolate; inside usually very deep brown with a lighter border. Apex blackish, sometimes having snowy dots or two crescents, the horns of one directed forward, of the other backward. Length 20, breadth 11, alt. 12 mill.

Sitka south to San Diego, California.

Patella insessa HINDS, Ann. and Mag. N. H. x, p. 82, t. 6, f. 3.—
Nacella insessa CPR., Suppl. Rep. Brit. Asso. 1863, p. 650.—*Acmæa insessa* DALL, Amer. Journ. Conch. vi, p. 244, t. 14, f. 3 (dentition).

A small, smooth, dark species. Faint, almost obsolete, radiating lines are usually perceptible. It is much larger than *A. paleacea* or *depicta* and is not so narrow. Compared with *A. asmi* it is larger, longer, and brown instead of black.

The Bay of Monterey is probably the central point for this species, in regard to numbers of individuals. It lives on the fronds of seaweeds.

A. ASMI Middendorff. Pl. 6, figs. 38, 39.

Shell small, thin but strong and solid, elevated, conical, the base short-oval, apex erect, a little in front of the middle; slopes of the cone somewhat convex. Surface lusterless, usually corroded, smooth except for very fine radiating striæ visible with the aid of a lens, but obsolete in adult shells. Color rusty black.

Inside black, with a brown zone just outside the muscle-scar.

Length 10, breadth 8½, alt. 7 mill.

Length 8½, breadth 7, alt. 8 mill.

Sitka to Turtle Bay, Lower California.

Patella asmi MIDD., Mal. Ross. ii, p. 39, t. 1, f. 5.—*Acmæa asmi* CPR., Amer. Journ. Conch. ii, p. 341.—*Collisella asmi* DALL, *l. c.*, vi, p. 252, t. 14, f. 7 (dentition).

In the suite of thirty or more of this species before me, no specimens show characters which warrant a union with any of the other species. It is an erectly conical, solid little shell, of a more rounded outline than *A. insessa*, and black instead of corneous in color. It is generally found living on *Chlorostoma funebrale* or other black shells.

A. DEPICTA Hinds. Pl. 6, figs. 40, 41.

Shell small, thin, long and narrow, the sides parallel; apex at the anterior fourth or third. Surface smooth, shining, having light growth-lines. Very light brown, with narrow dark brown stripes radiating from the apex and from the ridge of the back, where they form a series of v's.

Inside bluish-white, showing the color pattern of the outside through the shell. Length 11, breadth 4, alt. 3 mill.

Santa Barbara, Monterey, San Diego, California.

Patelloida depicta HINDS, Ann. and Mag. N. H. x, p. 82, t. 6, f. 4, 1842.—*Collisella ? depicta* DALL, Amer. Journ. Conch. vi, p. 254.

Resembles no species but *A. paleacea.* The long narrow form is caused by growth on *Zostera* fronds. There is some variation in the coloring, broad bands sometimes replacing the narrow lines. The front end is generally of a decidedly darker shade. A specimen of average proportions is figured.

The form of this species as developed when growing on a flat surface instead of a narrow frond, is shown in fig. 41 of pl. 6 drawn from a specimen in the Smithsonian collection. It measures, length 6, breadth 4½, alt. 1½ mill. Sculpture and coloration are as in the type.

A. PALEACEA Gould. Pl. 6, fig. 42.

Shell small, thin, long and narrow, parallel sided; apex near the front end. Surface sculptured with close radiating riblets. Color yellowish-brown, darker toward the margins and on the front slope. Length 7, breadth 1½, alt. 2 mill.

Monterey, Santa Barbara and San Diego, California.

A. paleacea GLD., Mex. and Cal. Shells p. 3, t. 14, f. 5.—CPR., P. Z. S. 1856, no. 40.—*Collisella paleacea* DALL, Amer. Journ. Conch. vi, p. 253 (dentition).

Smaller and narrower than *A. depicta,* the surface radiately ribbed and not variegated. The sculpture is quite distinct under a lens of moderate power.

A. TRIANGULARIS Carpenter. Pl. 7, figs. 74–78.

Shell small, rather thin, either oval or narrow and parallel-sided; elevated, apex subcentral, a trifle recurved; surface nearly smooth, but showing very fine radiating striæ under a lens. Color pure white, sometimes immaculate, but usually having 6 or 7 wide brown rays, which usually do not extend to either apex or basal margin. There is almost always a brown spot just behind the apex. Inside pure white.

Monterey and Baulinas Bay, California.

Nacella (? paleacea, var.) triangularis CPR. Proc. Cal. Acad. Sci. iii, p. 213, 1866.—*Collisella (?) triangularis* DALL, Am. Jour. Conch. vi, p. 254.—*Nacella casta* CPR. *olim.*

"They present nearly every variation in form, from wide, oval and nearly flat, to narrow, triangular, high and very compressed. The extreme apex is almost always black. It is usually furnished with a few dark brown stripes, radiating from near the apex but seldom reaching the margin in adult specimens. These however are wanting in some specimens. In all its forms it is a well marked species and cannot be united with any now known from the California coast. Dr. Carpenter proposes to rename this form specifically " *casta* " and to apply the term *triangularis* to the compressed variety only ; it is doubtful, however, if such a course would be admissible, as every transition in form can be observed in a very few specimens. (*Dall.*)

Typical *triangularis* is shown in figs. 77, 78. An example measures: length 6?, breadth 3, alt. 4 mill.

The form called CASTA is illustrated by figs. 74, 75, 76. This is really the *normal* form of the species, the other being modified by the narrow frond supporting it. A large example measures: length 12, breadth 9, alt. 4½ mill.

Var. ORCUTTI Pilsbry. Pl. 42, figs. 84, 85, 86.

Has the oval base of *A. casta*, but the apex is decidedly anterior, as in *A. persona*. Surface lusterless, having rather rude growthlines and *very obsolete*, low, wide radiating riblets, some at wide intervals slightly more prominent. Color white, or tinged with cream or green ; apex obtuse, eroded, but around the eroded area there are brown dots, indicating that the young were marked like *A. casta*. Interior white or fleshy-cream tinted, sometimes with slight greenish or brown stains in the cavity ; border wide, darker than the rest of the interior. Length 11¼, breadth 9, alt. 5¾ mill.

San Diego, California.

Specimens of this curious variety were received from Mr. C. R. Orcutt. It has the coloration of some specimens of *A. triangularis*, but the form recalls *A. persona*.

A. ROSACEA Carpenter. Pl. 7, figs. 71, 72, 73.

Shell small, conical, thin, smooth or with very obsolete ribs. The young are pale roseate, with few white and brown subradiating spots ; the adults have rosy brown and whitish streaks or are dotted with pale rose. Apex elevated, a little anterior ; inside white or rosy. Length 8, breadth 6⅓, alt. 3½ mill.

San Diego to Monterey, California.

Acmæa (*? pileolus var.*) *rosacea* CPR., Proc. Cal. Acad. iii, p. 213 ; Amer. Journ. Conch. ii, p. 341.—*Collisella* (*?*) *rosacea* DALL. Amer. Journ. Conch. vi, p. 256.

The shell is small, obtusely conical with an erect, subcentral apex. The ground color of the surface is a translucent white, suffused with rose toward the margin, where several indistinct rays of rose color appear. These are more evident on the inside. The extreme nucleus is usually white. The apex is profusely dotted with minute dark brown and opaque white specks of color, which are not rays, nor are they often arranged with any regularity ; these are more numerous on the posterior portion of the shell, but vary exceedingly, from a dark reticulated brown network of lines to wavy irregular penciling or sparse brown dots, usually most plenty on the interspaces of the ribs. The surface is smooth, especially in front, but from the apex radiate (especially on the posterior half of the shell) a number of very marked riblets which appear as if indented from below, and do not materially interrupt the smoothness of the surface, though the margin is rendered slightly crenulate by them. They are also of a more opaque white than the remainder of the shell, and sometimes form conspicuous white rays. (*Dall.*)

A. SYBARITICA Dall. Pl. 9, figs. 22, 23, 24.

Shell depressed, thin ; apex subcentral, more anterior in the young. General shape rounded-oval, hardly more narrow before than behind. Surface nearly smooth, with rounded concentric lines of growth, in young specimens a few faint hardly noticeable elevated radiating lines or riblets may be observed near the margin, which is entire. Internally smooth, border polished and also the cavity of the apex above the muscular impressions. Color a clear rose-pink, varying from quite deep and a little livid in some specimens, especially the young, to a very faint pink. Apex white, even in very young specimens entirely eroded, rather blunt and inconspicuous ; sides of the shell ornamented with rays of a darker shade of pink, more or less gathered in groups, and more or less evident, according to the shade of the remainder of the shell. Internally the visceral area is bluish-white, usually washed with a faint yellowish-brown, often hardly evident, in which case the area is whitish ; the successive layers of brown sometimes appear externally around the apex when eroded. The inner margin, and to some extent the whole interior, exhibit the external markings or rays through the somewhat pellucid shell. Texture hard and brittle. Epidermis exceedingly thin, usually evanescent; translucent, brownish. (*Dall.*)

Pribiloff Is. to Hakodadi, Japan; Aleutian Is.; southeast to Chirikoff Island.

A. sybaritica DALL., Amer. Journ. Conch. vi, p. 257, t. 17, f. 34, 1871; Proc. U. S. Nat. Mus. i, p. 341.

A beautiful species. The largest specimens attain one inch in length, but those before me measure scarcely over 10 mill. It is always much depressed. Inhabits rather deep water.

A. PERAMABILIS Dall. Pl. 33, figs. 80, 81, 82.

Shell thin, delicate, ovate; externally of a uniform dark-rose-color, with a few scattered irregular blotches of light or dark-brown, nucleus pale. Within polished, bluish-white, with a chestnut-brown spectrum with sharply defined edges, outside of which for a short distance the white is unsullied, but further toward the margin in adult specimens, radiating brown blotches may be observed forming a more or less interrupted band around the shell, which is wanting in the young. The margin is of the same deep rose as the exterior. Shell moderately elevated, with the apex well marked, sub-acute and situated in the central third. Nucleus smooth, pale, sharply decurved with a chink beneath it, in front. Sculpture of fine, sharp, elevated threads which extend from the vertex to the margin without bifurcation. These are crossed by very fine sharp lines of growth slightly elevated.

Length 1·03 in. lat. 0·8, in. alt. 0·33 in. Posterior slope slightly arched. (*Dall.*)

Shumagin group of islands; Alaska Territory, on rocks near low water mark.

This lovely species has no relations with *A. sybaritica* Dall and *rosacea* Cpr., except those of color. The two latter are much smaller and the rose color is much lighter and differently disposed. Its nearest allies are some varieties of *A. patina,* in none of which have I observed any approach to the color of this species, and which have a different nucleus, and the sculpture in slender rounded riblets instead of sharp threads. The shell of *patina* is also in general much more solid and thick. The animal partakes of the rosy hue of the shell except the margin of the mantle which is furnished with brown dots. It belongs to the subgenus Collisella.

It is worthy of note that when there is a brown marking on the exterior, in the region of the sub-marginal internal mottled band, the latter is interrupted by a white space corresponding in size and width to the external marking. (*Dall.*)

A. (Collisella) peramabilis DALL., Proc. Cal. Acad. Sci. iv, p. 302, Dec. 17, 1872; Proc. U. S. Nat. Mus. i. p. 341, 1878.

My figures are drawn from a type specimen. It is a most beautiful shell, as delicate in coloring as a rose-petal.

A. APICINA Dall. Pl. 7, figs. 66, 67.

Shell small, conical, thin, rounded, more or less elevated; whitish or isabelline, the apex erect, buff; inside buff, whitish or brown, smooth; provided with subobsolete lines of growth outside. Length 6, breadth 5, alt. 4 mill. (*Dall.*)

Pribiloff Is. on the north; Aleutians from Amchitka eastward to the Shumagins, 0–70 fms.

A. (Collisella?) apicina DALL, Proc. U. S. Nat. Mus. i, p. 341, 1878.

Among other small shells obtained from time to time on the beach or in the dredge, occasional specimens occurred which at first were supposed to be the young of *A. mitra* or pale specimens of *A. sybaritica*. After eliminating some of the these, there remained, after careful study, a residue which do not appear to coincide in character with any described species. They are small, thin, conical, with a blunt erect apex marked by a light yellow spot, the rest of the exterior white or faintly yellowish, marked by obsolete lines of growth, smooth or nearly so but not polished. Within, fresh specimens are yellowish, whitish or orange-colored, and quite polished. The outside is almost always covered with Nullipore. The chief characters are the rounded base, regularly conical and yellow spotted apex, with a thinner shell than *A. mitra*. (*Dall.*)

My figures are drawn from one of the types in the Smithsonian Institution, no. 30787 of the museum register. It is allied to *A. mitra* and *A. virginea*.

A. MITRA Eschscholtz. Pl. 3, fig. 50.

Shell dull-white, aperture nearly circular, wider behind, in some young examples somewhat elongated, oval; form conical, apex erect, nearly central, blunt, smooth, posterior surface usually straight, but occasionally a little convex; exterior smooth, marked with very faint concentric lines of growth, devoid of epidermis; margin entire, polished, with a narrow semi-pellucid rim inside.

Internally smooth or furnished with grooves radiating from the apex more or less strongly marked. Muscular impressions deep, strong, horse-shoe-shaped, with the marks of the anterior ends of the adductors rounded and broader than the rest, connected by a slender impressed line marking the attachment of the mantle. Young shells are often furnished with irregular riblets more or less strong, many or few in number, radiating from the apex, but stronger towards the margin. Color dead-white inside and out, often livid or tinged a fine pink or pea green from Nullipore, never wax-yellow or horny-pellucid as in the normal state of *Scurria scurra*.

Length 35, breadth 31, alt. 23 mill.
Length 23, breadth 20, alt. 17 mill.

Aleutian Is. to San Diego, California.

A. mitra Esch. in Rathke, Zool. Atl. pt. v, p. 18, t. 23, f. 4.—Dall., Amer. Jour. Conch. vi, p. 241.—*A. mammillata* Esch., *l. c.* p. 18.—*A. marmorea* Esch., *l. c.* p. 19—*Scurria mitra* Gray, Adams, Carpenter *et al.*—*S. ? mitra* Dall, Amer. Jour. Conch. v, p. 149 (dentition)—*Lottia conica* Gld., (part) Moll. U. S. Expl. Exped. p. 346.—*S. mitra var. tenuisculpta* Cpr. Amer. Journ. Conch ii, p. 346. —*Scurria ? funiculata* Cpr. Proc. Cal. Acad. Sci. iii, p. 214, 1865 ; Brit. Asso. Rep. 1863, p. 650 ; Amer. Jour. Conch. ii, p. 347.

The shell of this conical white species is very distinct from the others inhabiting West America. It resembles *Scurria scurra* Less., of Chili in form, but has not the waxen-yellowish outer layer of that species. The largest specimen I have seen measures, length 45, breadth 41, alt. 28 mill.

Var. TENUISCULPTA Cpr. Sculptured with distant radiating striæ or lirulæ.

Var. FUNICULATA Cpr. Shell small, whitish, regularly conical, apex acute, elevated, a little in front of the middle ; sculptured with strong rounded riblets, sometimes a little nodulous ; sometimes single, sometimes gathered into two's and three's ; with wide interspaces in which intercalary riblets appear. Length 6, breadth 4½, alt. 3 mill.

Monterey, California.

A curious small shell, having the contour of *A. mitra,* but with strong, smooth, crowded unequal ribs. The measurements are from the type in the Smithsonian Institution.

A. FASCICULARIS Menke. Pl. 6, figs. 50, 51, 52, 53.

Shell rather thin, depressed, oval; surface closely radiately striated, the striæ low, often obsolete but indicated by light dark lines.

"The prevailing tints are a reddish-brown outside, more or less mottled or striped with white; inside a prevailing white, more or less penciled or fretted with brown, and a border, sometimes white with a tessellated penciling of brown; sometimes a delicate fawn shading into a pinkish or slightly greenish tinge, with or without penciling. The body mark is of a dark lustrous brown, or very light with a greenish tinge, or nearly absent. It is large for the size of the shell, more or less removed from the margin. In shape, *A. fascicularis* is much longer, and generally smaller than *discors*. The standard color of *A. mesoleuca* is green, of *A. fascicularis* red. In *A. mesoleuca* the markings are laid on with stripes and patches, in *A. fascicularis* with very fine penciling. In the latter the outline of the body mark is much more regularly gathered up into points with concave margins between, the points often making regular lines radiating from the center. The surface of *A. mesoleuca* is covered with granulose ribs with soft interstices and a very thin smooth epidermis; that of *A. fascicularis* is very much more finely marked, showing under the glass, smooth ribs with the interstices extremely finely cancellated with very close, slightly rugose concentric striæ, covered with an extremely thin, rather velvety epidermis. The surface of *A. fascicularis* is much more generally abraded; and as the young shells were not uncommon in the Spondylus and Chama washings, while not one was found of *A. mesoleuca*, it is presumed that their station is different. The apex is sometimes brown, sometimes white; and in the smallest specimen, ·035 by ·025, shows no trace of being spirally recurved. The young shells are known by their finely cancellated texture and delicately reddish penciling; and generally, by a white spot proceeding from the apex posteriorly bounded by red lines. In all stages it is thin, and very glossy within." (*Cpr.*)

Length 27, breadth 21, alt. 6 mill.

Mazatlan, and Gulf of California generally.

A. fascicularis MKE. Zeitschr. f. Mal. 1851, p. 38.—CPR. Mazat. Cat., p. 255.—DALL, Amer. Journ. Conch. vi, p. 253, t. 14, f. 11 (dentition).—*A. mutabilis* MKE, in part, l. c., p. 37.—*Patella opea* RVE., Conch. Icon., f. 79, teste CPR.

A beautifully penciled species, allied to *S. mesoleuca*. The synonymous *P. opea* of Reeve as represented on pl. 6, figs. 52, 53.

A. STRIGATELLA Carpenter. Pl. 7, figs. 83, 84, 85.

The shell is ovate, a little wider behind, elevated; apex at the front fourth of the length. Young with *excessively fine close radiating striæ* crossed by growth-lines, largely worn off in adult specimens. Apex very acute in young, eroded, dark brown and polished in old shells. Coloration : Marked with irregular, forking black stripes on a white ground, interspersed around the apex when not eroded, with dots and small narrow or needle-shaped white streaks. Inside bluish-white, with the central area indistinctly irregularly clouded with brown. Border wide, vividly tessellated with blackish-brown.

Length 19, breadth 14, alt. 7 mill.

Cape St. Lucas, L. California.

A. strigatella CPR., Ann. Mag. N. H. 3d. Ser. xiii, p. 474.—*A. strigillata* CPR., Suppl. Rep. 1863, p. 618.—*Collisella strigatella* DALL, Amer. Journ. Conch. vi, p. 253.

The figures and description are from a type in the Smithsonian Institution collection. The *A. filosa* of Carpenter of which I have seen the type, seems to be a variety of this, differing in sculpture and the more depressed form. The Carpenterian specimen of *strigatella* in the Philadelphia collection has the radiating stripes less regular, more anastomosing.

A. FILOSA Carpenter. Pl. 7, figs. 80, 81, 82.

Similar to *A. mesoleuca* in form and texture; but sculpture much more delicate; young shell smooth; then with delicate acute lirulæ, scarcely granulose, very distant, sometimes obsolete; interstices wide, smooth; thin, flat. Oval, subdiaphanous, blackish-brown radiately strigate or variously maculated with corneous. Inside livid or whitish, the colors of the outside showing through; border broad, acute. Length ·7, breadth ·56, alt. ·12 inch. (*Cpr.*)

Panama.

Lottia ? patina C. B. AD., Cat. Panam. Sh. no. 367.—*Acmæa (? floccata, var.) filosa* CPR., P. Z. S. 1865, p. 276.

In shape and texture, but not in color or sculpture, these shells resemble *A. fascicularis*; in the latter respects, *A. strigatella*. (*Cpr.*)

I have examined the type of this species. It has much the appearance of *A. strigatella*, but differs in sculpture, having the lirulæ very delicate and more widely separated. It measures, length 17?, breadth 14, alt. 2⅝ mill.

A. SUBROTÚNDATA Carpenter. Pl. 33, figs. 1, 2, 3.

Shell similar to *A. filosa*, but subrotund, more elevated, vertex subcentral; color more intense, the corneous lines closer, narrow; young shell paler, with two triangular rays posteriorly; inside callus livid, thinner. Length ·53, width ·45, alt. ·15 inch. (*Cpr.*)

Panama.

Lottia sp. ind. a, C. B. Ad., Panam. Cat. no. 368.—*Acmæa (? floccata, var.) subrotundata* Cpr., P. Z. S. 1865, p. 277.

I give figures of the type of this species, no. 15922 of the Smithsonian Institution collection. It is a subcircular shell, with the sub-acute, erect apex near the center. The surface has sub-obsolete narrow, separated radiating threads. If held toward the light, fine close unequal brown radiating lines are seen through it. The outside appears of a dingy brownish, with obscure lighter lines. The inside is bluish-white, with a chestnut spot in the cavity of the apex: border wide, dark, with close lines of dark brown. Length 13½, breadth 11½, alt. 4 mill.; distance of apex from front end, 5¾ mill.

In my opinion the *A. vernicosa* of Carpenter is a variety of this.

A. VERNICOSA Carpenter. Pl. 33, figs. 99.

Shell small, subrotund, depressed-conical, apex situated at the front two-fifths of the shell's length; whitish-green, ornamented here and there with a few reddish-brown streaks; sometimes with white rays; sculptured faintly with acute, radiating very distant lines, sometimes obsolete; inside livid, callous, generally with a white spatula; base subplanate, border narrow. (*Cpr.*)

Length 7½, breadth 5¾ alt. 2 mill.

Panama.

Lottia sp. ind. b, C. B. Ad., Panam. Sh. no. 369.—*Acmæa (? var.) vernicosa* Cpr., P. Z. S. 1865, p. 277.

Had this form been brought from the China Seas it might have been taken for *A. biradiata* Rv. From its solidity, however, its rough exterior, and its callous interior, it appears to be adult. It is barely possible that it may develop into *A. vespertina*. It differs from the young of *A. subrotundata* in being much thicker and less spotted with the green tint. (*Cpr.*)

To Carpenter's description and remarks I add figures from the type, in the Smithsonian collection. It is a small, yellowish form, very solid, with erect, *acute* apex. I regard it as a probable variety of *A. subrotundata* Cpr. With "*vespertina*" it has nothing to do.

A. MITELLA Menke. Pl. 6, figs. 46, 47, 48, 49.

Shell small, conical, oval, apex in front of the middle; slopes straight or nearly so, surface radiately sculptured with larger and smaller riblets; color greenish or gray, often encrusted with a white coralline. Edge scarcely crenulated.

Inside white, with margin regularly spotted with black; central tract marbled with brown, generally with a bluish-black spot near the center. Length 11, breadth 8, alt. 5 to 6 mill.

Mazatlan to the Gulf of California.

A. mitella MKE. Zeitschr. f. Mal. 1847, p. 187.—CPR. Mazat. Cat. p. 210—DALL, Am. Journ. Conch. vi. p. 253, t. 14, f. 9 (dentition).—*P. navicula* RVE. Conch. Icon. f. 130.

Small, conical, rather finely radiately ribbed, the inside border regularly spotted with black.

A. DISCORS Philippi. Pl. 9, fig. 3, 4, 5.

Shell oval, solid, depressed, apex near the anterior third; slopes gently convex. Surface generally encrusted; having about 10–15 strong radiating folds around the apex, usually lost by erosion in adult shells and wholly absent on the later growth of the shell, which is closely, finely radiately striated. Margin very finely crenulated, often dotted with black. Color greenish or grayish-white, with more or less distinct radiating black lines.

Inside white, often mottled with purple or brown. Muscle-scar distinctly impressed. Length 48, breadth 37, alt. 14 mill.

Cape St. Lucas and Mazatlan to Panama.

P. discors PHIL. Abbild., t. 2, f. 6.—RVE, Conch. Icon., f. 78.— CPR. Mazat. Cat., p. 201.

This species is peculiar in being puckered around the apex, when not eroded. It is usually encrusted with coralline, algæ, etc.

A. ATRATA Carpenter. Pl. 7, figs. 61, 62, 63, 64, 65.

Shell solid, oval, apex at the anterior third; sculptured with unequal, irregular ribs, the interstices narrow. On a whitish ground it has black stripes and lines, and the eroded apex is black.

Inside white; the border yellowish-gray, tessellated with square black spots, usually bifid; the central area callous, white, clouded with brown and faint livid-purplish tints. It is excavated anteriorly at the cavity of the apex. Muscle-scar distinct, rugose.

Length 31, breadth 24, alt. 14 mill.

The young (figs. 64, 65) are flatter, and have very close, acute, unequal ribs, sharply crenulating the margin.

Cape St. Lucas, L. California.

A. (? *var.*) *atrata* CPR. Ann. Mag. N. H. 3d Ser., xiii, p. 474.— *Collisella atrata* DALL, Amer. Jour. Conch. vi, p. 225, t. 14, f. 15, 15a, (dentition.)

A perfectly distinct species, allied to *A. pediculus*, but more elevated, with differently patterned interior. My description and figures are drawn from the types in the Smithsonian Institution collection.

A. PEDICULUS Philippi. Pl. 6, figs. 34, 35; Pl. 7, figs. 68, 69, 70.

Shell normally flat, oblong, solid, with ten stout rounded ribs projecting at the margins, of which two are in the axis of length with four on each side; ribs and interstices radiately striated; yellowish-white generally with more or less of black or brown tortoiseshell markings within, sometimes with black between the ribs. Sometimes the shell is more rounded and the ribs more angular, in which state it might be taken for the young of *P. mexicana.* Occasionally a few other intercalary ribs appear. In a very few unusually large specimens, the ribs are nearly obsolete at the margin and the shell is much lengthened. The body mark varies as usual; when plain it is gathered into points as in *P. discors.* The very young shells appear not to develop the ribs marginally, in which state they might be taken for the young of *P. discors.* The stout ribs of the adult shell however bear no analogy with the very finely marked surface of the latter with its curiously puckered circum-umbonal portion. With the young of *P. mexicana* it has much more close analogies. The largest specimens of *P. pediculus* however do not at all run into the smallest of *P. mexicana.* They have all the appearance of being old shells, with the margin narrow and the shape long and irregular, while *P. mexicana,* as it is traced upwards, displays a very wide semitranslucent margin, and a broad regular shape, with the ribs not rounded and prominent but simply giving an angular form to the shell. Even when very young, they are almost always incrusted with corallinous matter. (*Cpr.*)

West Mexico, Mazatlan to Acapulco.

P. pediculus PHIL. Zeitschr. f. Mal. 1846, p 21, No. 8.—CARPENTER, Mazat. Catal. p. 200.—*P. corrugata* RVE., Conch. Icon. f. 132,

1855—*Collisella pediculus* DALL, Amer. Journ. Conch. v i. p. 255, t. 15, f. 16 (dentition).

The ribs extend to the margin in this species; in *A. discors* they merely pucker the central part. Some specimens have the interior most beautifully variegated with rich brown, black and steel-blue. One of these, from the Academy collection is figured on pl. 7, figures 68–70. The figures on pl. 6 represent Reeves synonymous *P. corrugata.*

A. STIPULATA Reeve. Pl. 6, figs. 27–31.

Shell solid, ovate-oblong, depressed, apex in front of the middle; slopes convex; sculptured with about 10 principal radiating ribs, scarcely larger than the numerous wide and narrow, rounded intermediate ribs, but giving a somewhat angular outline to the base. Color "greenish-black, faintly rayed with a few whitish lines" or light green, blotched with dark olive.

Inside white tinged with green, muscle-scar impressed, narrow, white, the tract within it olive-green; border of shell light green, clouded with blue-green and irregularly dotted with black.

Length 27, breadth 18, alt. 6 mill.

Panama.

P. stipulata Rv., Conch. Icon. f. 117, 1855.

A solid, depressed species, with low, rounded ribs, the interior having an olive or brownish-green central area. A specimen in the collection of the Smithsonian Institution agrees with the one figured from the Philadelphia collection (figs. 29, 30, 31). The other figures, 27, 28, are from Reeve, and represent a slightly different coloring.

Undetermined West American species of Acmæa.

Patella (Acmæa ?) personoides Midd., Beitr. zu einer Mal. Rossica ii, p. 37, t. 1, f. 2, (=*A. ancyloides* Midd., Bull. Acad. Sci. St. Pétersb. vi) from Kenai Bay, north-west Coast America, is probably a form of *A. pelta* Esch.

Patella (Acmæa ?) æruginosa Midd., *l. c.*, p. 38, t. 1, f. 1, is perhaps a form of *A. patina.*

Lepeta puntarenæ Mörch (Mal. Bl. vii, p. 175) is perhaps an *Acmæa.* The description is as follows: Shell oval, apex elevated, excentric; inside milk-white, outside yellowish, closely decussated with thick, narrow radiating and concentric subequal lines, the intersections nodose; margin entire. Has the aspect of *Lepeta cœca* Müll. Length 6, breadth 4, alt. 2 mill.

Puntarenas (*W. coast Central America*). A single specimen.

Differs from *L. cœca* in the very coarse and close concentric lines, and the numerous finer radiating lines.

* _* *

III. SPECIES OF THE WESTERN COAST OF SOUTH AMERICA.

Of the following species, I have not seen *A. coffea* Rv., *exilis albescens* and *nisoria* Phil. Of the others, numerous specimens have been examined.

A. SCUTUM Orbigny.   Pl. 4, figs. 77, 78, 79, 80, 81.

Shell rounded-oval, conical, the apex directed forward, nearer the middle than the front margin. Front slope straight, back slope somewhat convex. Surface nearly smooth, but finely radiately striate, the striæ obsolete on some specimens. Color black, dotted or spotted more or less profusely with white. Edge of the shell smooth, even.

Inside white, with a broad black margin (often spotted with white), and a dark brown central tract, which is usually partially or wholly concealed in old individuals by a white layer. Muscle-scar inconspicuous. Length 40, breadth 35, alt. 15 mill., or less.

Peru to Sts. of Magellan.

A. scutum ORB. (not Esch.), Voy. Amér. Mérid., p. 479, t. 64, f. 8–10.—CARPENTER, Amer. Journ. Conch. ii, p. 335.—*L. punctata* GRAY, where?

Has some resemblance to black forms of *A. patina.* It is a solid, black species, usually rather conical, more or less dotted with white. Young specimens have the apex decidedly hooked forward, and the central spot dark and prominent. It is a true *Acmæa.*

A. VIRIDULA Lamarck.   Pl. 1, figs. 1, 2, 3, 4, 5, 6.

Shell rounded-oval, conical, or depressed, the apex in front of the middle. Surface having about 20 low, rather wide radiating ribs, and obscurely, finely striated all over the ribs and interstics. The color-pattern is a close, fine network of light-green on a white ground, the intervals between the ribs with larger spots of green, ribs lighter; adults more of an even gray all over. Edge of shell even, smooth, with a green or light green border within.

Inside pure white in old shells, but with an irregular green or brown central spot in the younger stages.

Length 70, breadth 60, alt. 23 mill.

Peru to Chili.

Patella viridula, LAM. An. s. Vert. vii, p. 539—DELESSERT, Recueil, t. 23, f. 2—RVE. Conch. Icon. f. 26 a, b, c.—*Lottia viridula* GLD., U. S. Expl. Exped. p. 353, t. 30, f. 459.—*Patella pretrei* ORB., Voy. Amér. Mérid. p. 481, t. 78, f. 15, 16.—? *P.* (*Acmæa ?*) *plana* PHIL. Abbild iii, t. 2, f. 3.—? *A. nisoria* PH., Zeitschr. f. Mal. 1846, p. 49; Abbild. t. 2, f. 8.

Differs from *Scurria zebrina* in the character of the painting, and in the sculpture. The *viridula* lacks the cordon of branchial papillæ possessed by the other species.

Orbigny's *P. pretrei* (Pl. 34, figs. 9, 10) is evidently a variety. *A. plana* Phil. may perhaps be the young of this species; see under *A. araucana. A. nisoria* probably belongs here also. The description is as follows.

Var. NISORIA Philippi. Pl. 4, figs. 82, 83, 84.

Shell solid; suborbicular, elevated-conical, radiately obscurely ribbed-striate; whitish, subtessellated with lines, flames and dots of brown; apex situated at the front two-fifths of the length. Inside white, often brown in the cavity, the border dotted with brown. Length 15, breadth 14, alt. 6 mill. The impressed radii are irregular, and sometimes there is a weak indication of about 16 radial ribs. The nearly circular form, the height and thickness of the shell, as well as the coloration, mark out this species from its allies. (*Phil.*)

A. CECILIANA Orbigny. Pl. 34, figs. 17, 18, 19, 20, 21, 14, 15, 16.

Shell solid, oval, elevated, the apex in front of the middle, sometimes decidedly anterior; posterior slope frequently decidedly arched. The surface has 13 to 16 strong elevated radiating ribs, those upon the posterior slope most strongly developed. Color grayish, with several black lines in each intercostal space. Edge of shell more or less strongly crenated by the ribs.

Inside soiled white, the margin articulated with black. The central area has five or six longitudinal black stripes on a white ground. Length 17, breadth 14, alt. 7 mill.

3   *Falkland Is.; Strait of Magellan; Chili.*

Patella ceciliana ORB., Voy. Amér. Mérid., p. 482, t. 81, f. 4–6.—
GAY, Hist. Chil. viii, p. 260, 1854.—*L. viridula* GLD., in part,
Exped. Atlas, f. 459c, 459d, 459e, and text, p. 354.—*P. monticula*
NUTTALL mss. teste Gld., *l. c.*

This shell has much resemblance in form to strongly sculptured
examples of *Acmæa persona* of California; and it is as variable as
that species. Some specimens are so elevated that the height of the
cone is equal to the breadth of its base, and others are comparatively
depressed. The curious streaks of the central area are very characteristic of typical specimens, but in old individuals this is often more
or less obscured. Dwarfed specimens before me measure, length 9,
width 8½, alt. 7 mill. They have strongly arched basal side-margins,
probably from growing pebbles or small gastropod shells.

Figures 17, 18, 19 are D'Orbigny's figures of *ceciliana.*

I rank the following described form as a variety:

Var. SUBPERSONA Pilsbry. Pl. 34, figs. 11, 12, 13.

Shell the shape of *Acmæa persona.* Apex curved forward, its
distance from front margin ⅓ to ½ the length of the shell. Anterior
ribs obsolete, ribs at the back and sides 10–12 in number, rounded,
sometimes irregular, some of them obsolete. White, with v-shaped
or irregularly triangular black markings, the black often predominating, and frequently finely speckled or netted with white dots.

Inside white with gray and black spotted margin and solid, dark
chestnut central area.

Valparaiso, Chili, southward.

A. VARIABILIS (Sowerby) Reeve. Pl. 34, figs. 1–8.

Shell short, oval, very much depressed, the apex a little behind
the anterior third. Slopes somewhat convex. Surface sculptured
with numerous low, close, subequal riblets; gray or of a light green
tint, radiately painted with black lines of varying widths in the
intercostal intervals; edge smooth, even, scarcely modified by the
ribs.

Inside smooth, the muscle-scar white (rarely dark in the young),
the central area black, conspicuously mottled with white; border
broad, grayish or greenish-white, articulated with black; tract
between border and muscle-scar white, chestnut or blackish-brown.

Length 32, breadth 26, alt. 5–7 mill.

Chili.

Lottia variabilis Sowerby (in part), Zool. Beechey's Voy. p. 147, t. 39, f. 5, not figs. 3, 4; 1839.—*Patella variabilis* Gray, Rve., Conch. Icon. t. 25, f. 63, 1855.—*P. penicillata* Rve., *l. c.*, f. 102.—*P. (Acmæa?) lineata* Phil., Abbild. iii, t. 2, f. 1 (1846).

This is unquestionably a distinct species, characterized by its depressed contour and mottled central area. The outside is usually much eroded. The black lines of the exterior are often interrupted, and in large specimens they are arranged in about a dozen broad but indistinct bands. The synonymy given above is unquestionable.

A. ARAUCANA Orbigny. Pl. 16, figs. 21, 22, 22.

Shell ovate, extremely depressed, costate, whitish, inside whitish, margin crenulated, brownish. Diam. 30, alt. 4 mill. (*Orb.*)

Valparaiso.

P. araucana Orb., Voy. Amér. Mérid., p. 482, t. 65, f. 4–6.— *Collisella araucana* Dall, Amer. Journ. Conch. vi, p. 257 (animal). —? *Patella plana* Phil., Abbild., Patella, t. 2, f. 3.—*P. plana* Rv., Conch. Icon., f. 133.

The specimens before me are longer than Orbigny's figured type, and the ribs are more separated. The shell is frequently distorted.

I am inclined to believe that *A. plana* of Philippi is a synonym or variety of *A. viridula;* but *A. plana* of Reeve is very probably a synonym of *araucana.* My material is too limited to enable me to decide this question.

A. COFFEA Reeve. Pl. 4, figs. 91, 88.

Shell ovate, convexly depressed, radiately densely ridge-striated, always very much eroded; brown-black within and without, finely denticulated at the margin. (*Rv.*)

Valparaiso.

P. coffea Rv., Conch. Icon., f. 139, 1855.

A. EXILIS Philippi. Pl. 4, figs. 89, 90, 91.

Shell minute, thin, elongate-elliptical, convex, smooth, white, painted with brown rays; apex at the front third.

Length 6, breadth 4, alt 2½ mill.

In size and form just like *Patelloidea elongata* Q. & G., and perhaps is only a variety of it. That species should be *netted* over a greenish-yellow ground, according to the description; but the illustration of it shows simple red rays upon a gray ground, the two posterior rays broader and darker! The present species has about 20 to 24 dark reddish-brown rays, mostly in pairs, the interstices often milk-white, and a brown apex. The inside is similarly colored. One specimen is flatter. (*Phil.*)

Cape Horn.

A. exilis Ph., Zeitschr. f. Mal. 1846, p. 50; Abbild. iii, t. 2, f. 6. Probably a young shell.

A. ALBESCENS Philippi. Pl. 4, figs. 85, 86, 87.

Shell small, thin, ovate-oblong, elliptical, conical, sculptured with close, little elevated, obsolete radiating ribs; whitish, sometimes radiated and spotted with brown; apex elevated, situated at the front third. Inside white, margins incumbent at the ends.

Length 10, width 7½, alt. 4 mill.

Six specimens are before me. In the young the acute apex is very prominent, and also the ribs, of which I count about 24, are much more distinct. With age they become more and more obsolete, and in old examples which have lost the apex by erosion, they often can not be recognized with certainty. Young examples are frequently pure white with yellowish apex; older ones have more or less brown markings. The border is pale brownish, and does not lie in a plane, for the sides are higher than the ends. The inside is white, or brownish when there is brown to be seen outside. (*Phil.*)

Central Chili.

A. albescens Ph., Zeitschr. f. Mal. 1846, p. 50; Abbild. iii, t. 2, f. 7.

This seems to be young.

* * *

IV. WEST INDIAN SPECIES.

Many specimens of all described West Indian Acmacidæ, except the doubtful *A. Antillarum* Sow., have been examined by the author.

The West Indian Acmæas are few in number and of small size. Their affinities are with the species of western Mexico. The synonymy is involved in great confusion. Of the four commoner species I have seen hundreds of specimens, and am still inclined to consider them quite distinct.

The *Acmæa hamillei* of Fischer (Journ. de Conchyl. v. p. 277, and J. de C. 1872, p. 145, Pl. 5, f. 6) is the same as *Scutellina antillarum* (Shutt.) Dall, and as the latter was until 1890 a mere *mss.* name, the species should hereafter be known as *Scutellina hamillei*.

The species may be briefly diagnosed as follows :—

A. punctulata Gm. Depressed or low-conic, the apex acute, conical, subcentral ; riblets irregular ; pinkish, usually dotted with red. Inside white with a narrow, pale margin, the spatula white or salmon colored.

A. candeana Orb. Depressed, low-convex, closely and finely radiately striate. Light yellowish with narrow brown lines usually gathered into rays. Inside with a wide border and central brown area.

A. carpenteri Pils. Moderately elevated ; closely evenly and finely, rather obsoletely radiately striated ; rayed with pinkish-purple on a light buff ground. Inside with a rather narrow border, spotted at the terminations of the rays. The central area more or less marked with brown.

A. antillarum Sowb. See under *A. candeana.*

A. onychina Gld. Depressed, solid, having low but rather wide ribs, the ribs and intervals closely, finely, radiately striate.

A. leucopleura Gm. Conical or depressed, the apex subcentral ; coarsely radiately ribbed, ribs white, intervals dark or spotted with dark brown or black ; inside with a very narrow border and brown or brown-outlined central tract.

A. cubensis Rve. Conical, solid, with fine radiating riblets ; closely marked with radiating black lines which frequently bifurcate and anastomose ; inside with a very narrow dotted border and a dark central area.

A. PUNCTULATA Gmelin. Pl. 5, figs. 99, 100, 1, 2, 3, 4, 5, 6, 11, 12, 13.

Shell wide-oval, depressed or depressed-conical, the apex acute and subcentral. Surface sculptured with rather rude, low separated riblets of which some (about every third or fourth rib) are larger. Color yellowish or delicate pink, marked with distinct red dots between the riblets.

There is apparently no epidermis. Often the surface is finely speckled all over with pink, and there is sometimes a blotch of carmine on each side of the apex. The inside is white ; the border is very narrow, irregular, translucent gray. There are often reddish-brown dots sparsely scattered along the border, which seem to be under the gray edge. The central tract is short, wide, and either white or of a light salmon color. Length 24, width 19, alt. 7 mill.

Bermuda and Key West, Florida, south to Guadeloupe and Vera Cruz, Mexico.

Patella punctulata GMELIN, Syst. Nat. xiii, p. 3705, no. 68, and again p. 3717, no. 132.—? *P. punctata* LAM., An. s. Vert. vi, p. 333, no. 34.—*P. puncturata* LAM., l. c., p. 333, no. 35.—RVE, Conch Icon., f. 122 a, b.—*Patella pustulata* HELBLING, Beiträge zur Kenntniss neuer und seltener Conchylien, in Abhandlungen einer Privatgesellschaft in Böhmen zur Aufnahme der Mathematik, der Vaterländischen Geschichte und der Naturgeschichte, vi, p. 110, t. 1, f. 12, Prag, 1779.—GMELIN, Syst. Nat. xiii, p. 3720, no. 147.—ARANGO, Fauna Mal. Cubana, p. 230.—*Patella cubaniana* ORB., Moll. Cuba ii, p. 199, t. 25, f. 4–6.—*P. confusa* and *P. pulcherrima* GUILDING.

A very variable shell, but easily known from its West Indian fellows. There are sometimes delicate raised laminæ in the direction of growth lines, making wide low scales over the ribs. The young are almost always high and conical, very different from the depressed adult form. A frequent variation is illustrated by fig. 5, representing a convex, dome-shaped variety. Specimens from Key West, Fla., are small, thin, and narrower than most from the Islands; they form the var. *pulcherrima* Guild., of authors. The Bermuda specimens are small and conical.

Figs. 11–13 represent the *cubaniana* of Orbigny.

A. CANDEANA Orbigny. Pl. 5, figs. 91, 92, 93, 94, 95.

Shell ovate, depressed, the apex in front of the middle, slopes gently convex; surface finely, regularly striated radiately; gray or dull buff, with radiating black lines, often gathered into 7 to 9 broad rays.

Inside white (or brown tinted), with a wide translucent-gray border usually closely articulated with brown lines; central area generally brown or brown and white marbled.

Length 25, breadth 21, alt. 8 mill.

Bahamas and West Florida to Tobago, West Indies.

P. Candeana ORB., Moll. Cuba ii, p. 199, t. 25, f. 1–3.—*A. candeana* DALL, Cat. Mar. Moll. s.-e. U. S. p. 156.—*Patella tenera* C. B. AD., Synopsis Conchyliorum Jamaicensium, etc., in Proc. Bost. Soc. N. H. ii, p. 8, 1845.—RVE. Conch. Icon. f. 104.—*Patella* (*Acmæa?*) *elegans* PHIL., Abbild. iii, p. 34 (Patella p. 6, t. 2, f. 2.)

More finely and regularly striated than *A. punctulata*, the apex less erect, more anterior. The color-pattern also is quite different.

There is considerable likeness between this species and *A. fasciculuris* Mke, a form from the Gulf of California.

In thin or young specimens the lines of the outside show through the bluish-white internal layer. Sometimes the light ground-color, sometimes the dark markings predominate in the coloration.

The form called *elegans* by Philippi (see pl. 5, figs. 96, 97, 98) differs in having the border very narrow. It was described from La Guayra, Venezuela, in 1846; and if I were assured of its identity with *candeana*, I would call the species *Acmœa elegans*, as that name has some years priority over Orbigny's.

A. candeana is more depressed than *A. Carpenteri*, with wider, differently marked internal border and different coloration.

Var. *antillarum* Sowb. A curious varietal form referable to *A. candeana* is figured on pl. 42, figs. 92–95. It is decidedly elevated, the surface having minute radiating striæ around the apex, becoming obsolete below. It is rayed with narrow, light blue stripes on a blackish ground, dotted with blue around the apex. Inside bluish-white, the body-mark deep chestnut in the young and half-grown (fig. 95), but partly overlaid with white in the adult; border dark, articulated with light. Length 20, breadth 16, alt. 7½ mill. The specimens were collected at St. Kitts by Dr. Wm. H. Rush, U. S. N.

I have little doubt that this is the *Lottia antillarum*, some specimens in Dr. Rush's collection agreeing exactly Sowerby's figure.

Lottia antillarum Sowerby (pl. 5, fig. 7), has never been characterized in any way. The original figures are copied on my plate. (see *Sowerby*, A Conchological Manual, p. 59, fig. 231, 1839.)

A. CARPENTERI Pilsbry. Pl. 33, figs. 70, 71, 72, 73.

Shell oval, rather thin; apex situated near the anterior third of the length, obtusely rounded; surface very finely, evenly radiately striated all over, rayed with purplish-pink on a light buff ground, the rays 8–12 in number, the anterior ones narrower; more or less flecked with light brown around the apex.

Inside white, either immaculate or having a brown spot in the cavity and a narrow outline around the central area; rays of the exterior generally indistinctly showing through; margin narrow, buff, with brown spots at the ends of the rays.

Length 20½, breadth 15½, alt. 8 mill.

West Indies.

A. melanosticta CARPENTER, in Mus. Smithsonian Institution, not *P. melanosticta* GMEL., Syst. Nat. xiii, p. 3724; founded on *Die weisse grau gestrahlte und schwartz punctirte Napfschnecke* of SCHROETER, Einleitung in die Conchylienkenntniss, ii, p. 497, t. vi, f. 9, an unidentifiable but certainly different *Acmæa*.

The young or half grown are often prettily flecked with light brown v-shaped or zigzag markings, most distinctly seen on the inside (fig. 73). The rays are sometimes absent, and are usually obscure on unworn adult specimens, which are dull and lusterless outside. My figures are drawn from the types in the Smithsonian Institution.

The striæ are more even, more obsolete than in *A. candeana;* the apex is more elevated, more anterior, the coloration is of a different pattern, and the internal border is narrower, and either unspotted or with fewer spots than in the *candeana.*

The synonymy given above is merely nominal, and is introduced only to explain the labels given by Carpenter to the specimens in the Smithsonian and probably in other collections. Anyone who suspects that this is the *melanosticta* of Gmelin, after a perusal of his description, may consult the reference in Schröter, and be satisfied that it is not. Gmelin's diagnosis was compiled from Schröter's.

A. LEUCOPLEURA Gmelin.   Pl. 5, figs. 16–26.

Shell solid, rounded-oval, erectly conical or depressed, the apex subcentral; surface sculptured with about 12 strong primary ribs, but large specimens have 20 to 30 ribs around the base, the number being increased by the interposition of interstitial ribs, not reaching to the apex. Ribs white, interstices brown or black.

Inside white with a narrow gray border usually black dotted, the central area brown or outlined with brown.

Length 23, breadth 18, alt. 9 mill.

Southwest Florida to Guadeloupe.

Patella leucopleura Gmel., Syst. Nat. xiii, p. 3699, no. 34.—LAM., An. s. Vert. vi, p. 332.—*Acmæa melanoleuca* DALL (not Gmelin), Catal. Mar. Moll. S.-E. U. S. p. 156.—*P. albicosta* C. B. AD., Proc. Bost. Soc. N. H. ii, p. 8, 1845.—*P. albicostata* RVE., Conch. Icon. f. 128, 1855.—*P. balanoides* RV., *l. c.*, f. 137.—*P. occidentalis* RV., *l. c.*, f. 135.—? *P. cimeliata* RV., *l. c.*, f. 116.

There is considerable variation in this species, but it is always much more coarsely ribbed than *A. cubensis* Rve. The extremes of form are shown by figs. 18 and 19, representing small specimens. The best figure referred to by Gmelin is that in Knorr's " *Vergnügen der Augen,* etc. vi, pl. 28, fig. 9, really an excellent figure. The figure in Lister is less characteristic; and Martini's figures do not belong to this species at all. This last fact precludes the use of the name *melanoleuca* Gm. for this species, and indeed it is only in the last few years that the name has been so used. The *melanoleuca* of Reeve is not this shell. Unquestionable synonyms are *P. albicosta* Ad. (*albicostata* Rv.), figures 22, 23; *P. balanoides* Rv. figs. 24, 25; and *P. occidentalis* Rv. fig. 26. A more doubtful form is *P. cimeliata* Rv. figs. 14, 15, of pl. 5, said to be from Honduras.

A label in our museum, written by Robert Swift, gives the name " *Patella fungus* Mke." as a synonym of *P. albicosta.* I have seen no description of *P. fungus.*

A. CUBENSIS Reeve. Pl. 4, figs. 56, 57, 58, 59, 60, 69, 70.

Shell solid, ovate, elevated, the apex a little in front of the middle; surface sculptured with numerous narrow riblets, often obsolete. Closely marked all over with bifurcating and anastomosing black lines on a white ground, the black sometimes confluent into large blotches. Edge of shell smooth.

Inside white with a brown central area marbled with white; rarely entirely white; border very narrow, black with light dots, or light with black dots. Length 21, breadth 16, alt. 10 mill.

St. Thomas, St. Croix, Guadeloupe etc., West Indies north to the Bahamas.

Patella cubensis RVE., Conch. Icon., f. 125, 1855.—*P. melanoleuca* RVE. *l. c.,* f. 134, not *P. melanoleuca* GMELIN.—*P. leucopleura* REEVE, *l. c., f.* 138.

This is a solid, conical species, having the riblets much more numerous and finer than *A. leucopleura,* and marked in a peculiar pattern of forking and anastomosing black lines. The variations of this pattern are sufficiently shown by the figures.

Of the names applied to this species, none prior to Reeve's can be identified with any confidence. *P. melanoleuca* Gm. is much more likely to be *Subemarginula notata* L. than this species. The figures in Martini referred to by Gmelin correspond exactly with half-grown *notata,* but are too depressed for *A. cubensis;* and in any case *leucopleura* has priority. *P. melanosticta* Gm. is a wholly different shell, evidently a depressed species ("*planiuscula*"), larger and differently marked. Neither description nor figure apply to *cubensis.*

Figs. 69, 70 are copies of Reeve's illustrations of his *cubensis*; figs. 56, 57 are *melanoleuca* Rv. not Gmel.; figs. 58, 59, 60 are drawn from a large specimen of typical *cubensis*.

Var. SIMPLEX Pilsbry.   Pl. 4, figs. 63, 64, 65, 66, 67, 68.

Shell having a tendency to be squarely orbicular when fully adult, and much elevated, the black lines continuous, separate, rarely splitting more than once.   This is the *P. leucopleura* of Rv., not Gm. nor Lam.   Specimens from the Bahamas were collected from the shells of *Livona pica*; figures 66, 67, 68 were drawn from one of these.   Fig. 63 represents a large specimen of the typical form of *simplex*.

Another modification of either the *cubensis* or the *leucopleura* stock is figured on pl. 4, figs. 61, 62.   It is small, conical, apex erect and subcentral.   When cleared of the extraneous calcareous coating it is seen to be either smooth or obsoletely ribbed, the ribs narrow, separated; marked with blackish-brown radiating blotches which split into two or three branches toward the basal margin. Inside with a dark central area with or without a white central spot, the border black spotted with white.   The external color-markings may be faintly seen inside.   The exact status and affinities of this form are somewhat doubtful.   It is abundant at Manchconiel and Port Antonio, Jamaica.

The locality "West Indies" given in the books for *Subemarginula notata* L. is probably owing to a confusion of that species with these similar Acmæas.   Specimens of *notata* marked "Ceylon" are before me, and I have no doubt that it is an oriental species.   The very strongly inflexed ends of the muscle-scar, and the slight trace of a groove between the apex of the cavity and the front margin will enable one to separate *S. notata* from all Docoglossate limpets.

A. ONYCHINA Gould.   Pl. 34, figs. 22, 23, 24.

Shell small, thin, irregular, depressed, broadly ovate, with twenty or more obsolete, unequal ribs.   The general color is ashy green, with deep olive stripes between the ribs.   Besides this, the whole surface is marked with fine radiating striæ, and by crowded loose lines of growth, giving the surface a decomposing aspect.   Margin sharp, irregular; interior mottled with cream-color and clear chestnut-brown; central spatula thick and well marked; a marginal border is striped alternately chestnut and yellow, answering to the

ribs and intervening spaces. Summit eroded, dark brown. (*Gld.*)
Length 20, alt. 6 mill.

Barbados to Rio Janeiro, Brazil.

? Acmæa subrugosa ORB., Voy. Amér. Mérid., p. 479. 1847.—*Lottia
onychina* GLD., U. S. Expl. Exped., p. 355, atlas, f. 461.—*Collisella
subrugosa* DALL, Amer. Journ. Conch. vi, p. 255, t. 14, f. 14 (dentition).—*Patella (Tectura ?) Mülleri* DKR., Jahrb. D. M. Ges. ii, p.
246, 1875.

The specimens before me are much eroded. I have no doubt that
P. mülleri Dkr. is synonymous, but give the original description
here for comparison. *A. subrugosa* Orb. is also in all probability
the same, but the description is wholly insufficient for identification,
and should not be allowed to displace Gould's excellent diagnosis
and good figures.

P. mülleri. Shell solid, ovate or elliptical, often irregular, more
or less convex, seldom conic; with weak ribs and fine striæ, which
also cover the ribs, but are seen only on fresh, uncorroded specimens.
The obtuse apex lies at about two-fifths the shell's length. From it
unequal dark brown streaks radiate, which often split, and which
show through on the shining inside, and are especially distinct on
the border. The center is whitish or liver-brown; margin acute,
simple or subcrenulated. Length 22–24, alt. 10 mill.

*Bay of Maldonado, and Destero, Province of Sta. Catharina,
Brazil*, abundant.

Undetermined West Indian species.

Patella (Acmæa ?) antillarum Philippi (as of Sowb.) Pl. 5, figs. 8,
9, 10.

This is a dark colored, very finely striated, depressed shell. I
have seen nothing like it from the West Indies. Philippi gives
Antilles as the habitat. There is no resemblance between this and
antillarum Sowb. Compare *Scurria parastica* Orb.

(See *Phil.*, Abbild., *etc.* iii, Patella, t. 2, f. 12.)

* * *

V. JAPANESE SPECIES.

The species of Japan are allied to those of Western North
America. Specimens of all of them have been examined by the
writer.

A. SCHRENCKII Lischke. Pl. 2, figs. 21, 22, 23, 24.

Shell elliptical, much depressed, rather thin. The apex is situated between the front sixth and eighth of the shell's length; posterior slope convex, its curve generally higher than the apex. Sculptured with very fine, unequal thread-like riblets, which are very closely granose; of an olive-ashen color, variously marbled with blackish-olive.

Inside light blue with a broad blackish border and an ill-defined dark chestnut central area.

Length 31, breadth 22–24, alt. 4–6 mill.

Ojima and Nagasaki, Japan.

P. schrenckii LISCHKE, Mal. Blat. xv, p. 220; Jap. Meeres-Conchyl. i. p. 107, t. 8, f. 1–4.—*Tectura schrenckii* DKR., Ind. Moll. Mar. Jap., p. 155.

My description is drawn from typical specimens collected at Ojima by Mr. Frederick Stearns. The species is variable in markings, the mottling sometimes being converted into stripes obliquely radiating and curving from the central dorsal region. It is closely allied to *P. concinna*, but is of a longer elliptical outline, is more depressed, and the granulation is finer. Still, I am disposed to believe that the two will be united when fuller collections are made.

A. CONCINNA Lischke. Pl. 2, figs. 12, 13, 14, 15, 16, 17.

Shell oval, rather thin; apex at the front fifth or sixth of the shell's length; posterior slope convex. Surface sculptured with close fine unequal riblets, finely and distinctly granulose; of a uniform blackish-olive shade, or variegated with olive, green or brown on a very light green ground.

Inside light blue, with a wide dark or spotted border, and an ill-defined central darker tract. Length 24, width 20, alt. 6 mill.

Yokohama to Enoshima, Japan.

P. concinna Lischke, Mal. Bl. xvii, p. 25; Jap. Meeres-Conchyl. ii, p. 98, t. 6, f. 1–6.—*Tectura concinna* DKR., Ind. Moll. Mar. Jap. p. 154.—*P. granostriata* SCHRENCK, Amurl. Moll. t. 14, f. 1–3.

Closely allied to *A. schrenckii*, probably a variety of that species, but rounder, more elevated, more distinctly granulose. These Japanese forms have much in common with *A. scabra* of the Californian coast, but they are abundantly distinct in sculpture, the position of the apex, and the general tone of coloring.

Numerous specimens collected by Frederick Stearns and others are before me, exhibiting considerable variation. A variegated specimen is figured on pl. 9, fig. 30. This species is the *P. grano-striata* of Schrenck. The true *granostriata* of Reeve (Conch. Icon., f. 126) described without habitat, is very likely the same as *concinna;* but without more information it would be mischievous to use that name for this or any other species.

A. HEROLDI Dunker. Pl. 2, figs. 18, 19, 20, (enlarged); Pl. 9, figs. 17, 18, 19, 20, 21.

Shell small, ovate-elliptical, not much elevated, sculptured with more or less distinct close riblets; whitish, ornamented in various patterns with brown; vertex elevated, situated at the anterior fourth of the shell's length. Inside white or bluish, brown in the middle. Length 11, width 8, alt. 4 mill. (*Dkr.*)

Nagasaki and Decima, Japan.

P. heroldi DKR. Moll. Jap. p. 24, t. 3, f. 13.—*Acmæa heroldi* LISCHKE, Jap. Meeres-Conchyl. ii. p. 96.—*P. conulus* DKR. *l.c.* p. 24, t. 3. f. 19.—*P. pygmæa* DKR. *l. c.* p. 24, t. 3, f. 20.

A small species, varying much in degree of elevation, coloration and position of the apex. It passes through the same mutations as most other Acmæas. I have not seen the typical form.

Form *conulus* Dkr. Pl. 9, figs. 17, 18.

Small, much elevated, apex more central. Length 8, breadth 6½, alt. 5½ mill. The figures are drawn from a specimen before me. Of this form I have seen a good many examples. The radiating riblets are wanting or obsolete; most are marked with dull olive or brown stripes.

Form *pygmæa* Dkr. Pl. 9, figs. 19, 20, 21 (enlarged).

This is the narrow form, probably growing on fuci. It is very finely striated radially. The size is smaller than the figures.

A. DORSUOSA Gould. Pl. 33, figs. 88, 89, 90; Pl. 9, figs. 15, 16.

Shell oval, elevated conical, having the form of *A. persona* Esch.; solid. Apex curved forward, situated one-eighth of the shell's length from the anterior end. Surface having strong irregularly nodose radiating cords, separated by spaces of greater width than the ribs and having occasional strongly marked concentric depressions, indicating periods of growth arrest. Color, blackish-brown with irregular obscure lighter olive-yellow streaks in the interstices. Interior bluish-white; border not distinctly defined, having indistinct dark tesselations. Central area defined and obscurely clouded with dark chestnut. Length 30, breadth 24½, alt. 14 mill.

Hakodadi, Japan.

A. dorsuosa GLD., Proc. Bost. Soc. N. H. vii, p. 162,1859. Otia, p. 115.—WATSON, Challenger Gastrop. p. 29, t. 4, f. 1.

This species, which I have figured and described from types, is evidently allied to *A. persona* and *pelta* of the Californian coast, but is quite distinct. The ribs are variable in number, 20 to 27 being developed; they are obsolete in front. The periods of arrested growth are strongly marked. It is likely that "Tectura digitalis" reported from Hakodadi by Dunker (Ind. Moll. Mar. Jap. p. 154) is this species.

No other Japanese *Acmæa* is so strongly ribbed as this.

Watson figures a slightly differing variety from Oosima. See pl. 9, figs. 15, 16.

* * *

VI. INDO-PACIFIC, AUSTRALIAN AND NEW ZEALAND SPECIES.

The Australian region is rich in forms of *Acmæa*, but the number of species is doubtless less than are herein admitted. The lack of extensive suites has prevented me from working up the synonymy as fully as I would desire.

There are but few Polynesian Acmæas, and a still smaller number are found in the Indian Ocean.

The following species I have not seen: *A. biradiata* Rv., *rugosa* Q. & G., *cingulata* Hutton, *conoidea* Q. & G., *petterdi* T.–W., *septiformis* Q. & G., *laciniata* Rv., *cantharus* Rv., *scabrilirata* Ang., *flammea* Q. & G., *subundulata* Ang., *pileopsis* Q. & G., *orbicularis* Q. & G., *mixta* Rv., *elongata* Q. & G., *punctata* Q. & G., *squamosa* Q. & G.

A. CONOIDALIS Pease. Pl. 33, figs. 93, 94, 95.

Shell solid, thick, short-oval, elevated, straightly conical, the apex erect, pointed, subcentral, slopes of the cone straight or the posterior a little convex. Surface sculptured with numerous close, rather fine radiating threads, alternately larger. Whitish or yellowish-white. Edge thick, very finely crenulated.

Inside yellowish, the central area opaque white. Basal side margins somewhat arched. Length 22, breadth 17, alt. 11 mill.

Raratonga, Cook Is.; Hervey Is.

Tectura conoidalis Pse., Amer. Journ. Conch. iv, p. 98, t. 11, f. 22.—Martens & Langkavel, Donum Bismarkianum, p. 51.

Allied to *A. mitra*, but more obviously radiately ribbed than any of the varieties of that species, and differing in the yellowish interior, with opaque white central tract. The largest specimen before me measures, length 28, breadth 23, alt. 14 mill. The species varies considerably in degree of elevation.

A. GARRETTII Pilsbry. Pl. 33, figs. 96, 97, 98.

Shell small, oval, conical, the apex acute, erect, slightly in front of the middle; slopes convex. Surface having numerous close radiating riblets, of which a few (about seven) are generally larger. Color dull blackish-brown, or gray with lines of dark.

Inside white, the large central area either solid blackish-brown or clouded; border narrow, black or light, dotted dark brown.

Length 13, breadth 10, alt. 5 mill.

Length 11, breadth 9, alt 5½ mill.

Viti Islands.

A small species, collected by Garrett, and apparently distinct from the described Australasian forms, as well as from the few described from Polynesia. The figures represent a strongly sculptured specimen, but some specimens have the riblets more nearly equal in size, and in some the color, both inside and out, is lighter. It is a very solid little shell, somewhat similar to some forms of *A. cubensis*. It is much more solid and differently sculptured than *A. mitella*.

The only species reported from the Viti Islands besides *A. saccharina* is a *Patella nana* Dkr., enumerated in the Mus. Godeffroy Catal. v, p. 149, 1874; but I have been able to find no description of it. It is again catalogued in the Catal. vi, Nachträge zu Catal. v, p. 91, 1877.

A. STRIATA Quoy & Gaimard. Pl. 19, figs. 37, 38; Pl. 35, figs. 25–29.

Shell orbicular, convex, very delicately striated, brown or black; inside sky-blue, the margin brown. Apex obtuse.

A large species, nearly round; the summit is anterior. Length 30, breadth 25, alt. 6 mill. (*Q.*)

Licoupang, Celebes.

Patelloida striata Q. & G., Astrol. p. 353, t. 71, f. 8–11.—*Patella striata* Rv., Conch. Icon. f. 58 (not fig. 99).

The additional locality, *Philippine Is.*, is given by Reeve.

A number of specimens of this form are before me. They differ from *A. borneensis* Rv. only in being more distinctly striated and darker colored. The largest specimen I have seen measures, length 43, breadth 39, alt. 9 mill.

VAR. BORNEENSIS Reeve.   Pl. 35, figs. 25, 26, 27, 28, 29.

Shell rounded-oval, rather thin, depressed, the apex at about the anterior third; slopes convex; surface nearly smooth, but obscurely radiately striated; dull-grayish-white with close hair-like radiating black lines and more or less speckled with brown or blackish, these last markings frequently forming obscure darker rays; sometimes blackish all over.

Inside bluish-white, with a wide dark border and a small brown central area. Muscle-scar inconspicuous, remote from the margin.

Length 29, width 24, alt. 7 mill.

Amboyna; North Coast of Borneo.

P. borneensis Rv. Conch. Icon. f. 113, 1855.—*Collisella borneensis* DALL., Amer. Journ. Conch. vi, p. 259, t. 15, f. 17 (dentition); t. 17, f. 38 (shell).—*A. Bickmorei* DALL *mss.*

A depressed species, nearly smooth, hair-lined with deep brown or black and more or less speckled with the same color. I have seen none with the rays so pronounced as in Reeves figure.

A. FLUVIATILIS Blanford.   Pl. 35, figs. 40, 41, 42.

Shell much depressed, conical, subcircularly oval, thin, covered with a very dark olive epidermis, always eroded at the subcentral apex; marked with fine radiating raised lines, and with close and minute concentric striæ of growth.

Inside the shell is bluish-white, sometimes with one or more milky concentric bands, or the whole interior is milky, except the cavity of the apex which is invariably ferruginous.   (*Blanf.*)

Length 21½, breadth 20, alt. 6 mill.
Length 20, breadth 17, alt. 5½ mill.
Length 14, breadth 12, alt. 4 mill.

Delta of the Irawadi River, Pegu.

Tectura fluviatilis W. T. BLANFORD, Journ. Asiatic Soc. Bengal, xxxvi, pt. 2, 1867, p. 62, t. 13, f. 2, 3, 4 (1868).

This species is found on rocks, rarely on trunks of trees, in many of the creeks near high-water mark, in brackish water. It was not met with near the sea, where the water was very salt. The foot is large, filling the cavity of the shell, muzzle broad, tentacles long and fine, mouth not notched beneath. It does not appear to keep to one place and form a hole for itself like some Patellæ, but it is very sluggish in its movements. (*Blanf.*)

The only specimen of this species which I have seen is in the collection of Dr. W. H. Dall. It has the appearance of an ordinary *Acmæa* of the type of *A. striata* Q. & G.

A. BIRADIATA Reeve. Pl. 42, figs. 88, 89.

Shell nearly orbicular, conoidly depressed, apex nearly central; radiately striately ridged; whitish, rayed with blue-black near the margin, conspicuously ornamented posteriorly with two broad, pale bands; interior vividly painted with the same color. (*Reeve*).

China Seas.

P. biradiata RVE., Conch. Icon. f. 142.

A. HIEROGLYPHICA Dall. Pl. 33, figs. 77, 78, 79.

Shell small, stout, rugged, with a sub-central, more or less eroded apex moderately elevated. Muscular impression pyriform. Shape of shell ovate; exterior with rather strong white ribs, 14–20 in number, with riblets between them, interspaces brown. Striæ of growth somewhat imbricated, less prominent on the ribs. Internally white, with brown maculæ on the margin corresponding to the brown interspaces of the exterior. Margin strongly crenulated. Spectrum pyriform with the smaller end anterior, consisting of a sharp black line forming a pyriform figure with three longitudinal black lines inside of it. In the larger specimens these have a faint bluish halo about them, but in the smaller they are simply black on a white ground. The same figure of less size is conspicuous on the outside of the eroded apex, soft parts unknown. (*Dall.*)

Length 7, breadth 5¾, alt. 1½ mill.

China.

A. hieroglyphica DALL, A. J. C. vi, p. 258, t. 17, f. 37.

This little species of the *A. spectrum* group, has much resemblance to the Chilian species *A. ceciliana*. I have seen no specimen but the type, which was found in a box of Chinese shells in a San Francisco tea shop.

A. SACCHARINA Linné. Pl. 36, figs. 60, 61, 62, 78; pl. 18, figs. 31, 32; pl. 24, figs. 12, 13.

4

Shell solid, conical, having seven prominent, angular radiating ribs, projecting at the margin ; surface finely radiately striated when not eroded ; variously marked with black blotches, usually v-shaped, on a whitish ground.

Inside white, the border gray marked with black or entirely black ; central area with a patch of black or deep brown, covered over with white in old specimens. Length 40, breadth 30, alt. 18 mill.

Viti and Fiji Is ; Amboyna to Japan.

P. saccharina L., Syst. Nat. x, p. 781, no. 660.—REEVE, Conch. Icon. f. 72.—LISCHKE, Jap. Meeres-Conchyl. p. 113, 1869.— DUNKER, Index Moll. Mar. Jap. p. 155.—*Acmæa saccharina* HANLEY, in Wood's Index Test., 2d edit. p. 185, 1856.—*Collisella saccharina* DALL, Am. Journ. Conch. vi, p. 259, t. 15, f. 18 (dentition).— *Patella lanx* Rv., Conch. Icon. f. 82.

The typical *saccharina* is best represented by the figures 60–62 of Pl. 36. Sometimes smaller ribs are developed between the principal ones, as in Pl. 18, figs. 31, 32. Reeve's *P. lanx* (Pl. 24, figs. 12, 13,) is not a constant color variety.

Var. STELLARIS Q. & G.   Pl. 36, figs. 63, 64, 67, 68.

Resembles *saccharina* in form. Central area of the inside dotted with brown.

New Ireland, etc.

Patelloida stellaris Q. & G.. Voy. Astrol. Zool. iii, p. 356, t. 71, f. 1–4, 1834—*Patella stella* LESSON, Voy. de la Coquille, Zool. ii, p. 421, 1830.—*P. octoradiata* HUTTON. See below.

A beautiful variety, of which numerous specimens from the collection of Mr. John Ford, of Philadelphia are before me, (figs. 63, 64).

The types of Quoy & Gaimard (figs. 67, 68,) differ somewhat from the shells before me, in having the ribs more prolonged at the margins, but agree in the number of principal rays, *seven,* and the dotted spatula. Reeves *P. stellaris* is unlike the true *stellaris* in having eight principal ribs and a solid brown spatula. See below.

The name proposed by Lesson is anterior in date (1830), but being briefly described without a figure, is scarcely entitled to displace the unmistakable diagnosis and good figures given by Quoy.

Var. PERPLEXA Pilsbry.   Pl. 36, figs. 69, 70, 71.

Depressed, star-shaped, with four large rounded ribs behind, three in front of the apex; ribs and intervals closely striated. White or creamy, with fine radiating reddish-brown lines.

Interior white with a few flesh-colored spots; border narrow, dotted with pinkish. Length 31, breadth 27, alt. 8 mill.; a smaller specimen measures, length 21, breadth 20, alt. 4½ mill.

Australia (Phil. Acad. Coll.)

? *P. stellaris* REEVE, Conch. Icon. f. 114, not *stellaris* Q. &. G.— ? *P. octoradiata* HUTTON, Cat. Mar. Moll. N. Z. p. 44, 1873.—? *P. stellaris* (Q. & G.) HUTTON, Journ. de Conchyl. 1878, p. 37.

It is with great doubt that I give the above synonymy for this species or variety, for Hutton has never made clear what his *octoradiata* really is; referring it first to *stellaris* Q., and then omitting it from his Manual of 1880. It is, however, fair to assume that in 1880 he included his specimen under the name *P. stella* Lesson. Lesson's species is no doubt completely synonymous with Quoy's *P. stellaris;* and I, therefore, offer a new and definite appellation for this form. The *Patella stellaris* Rv. (not Q. & G.) is an octoradiate limpet resembling somewhat the shells above described, but not agreeing in characters with any of the specimens before me. Reeve's figures are copied on pl. 36, figs. 65, 66. They should be compared with certain varieties of *Patella pentagona.*

Under whatever name this variety or species is ultimately placed, it is well characterized by the number of ribs, which is constant in the large suite before me, the rounded form of the ribs and their striated surface. The interior does not have the dots characteristic of *A. stellaris* Q. & G. The ribs are not carinated as in *Patella longicosta* Linn., nor are interstitial riblets developed. Two ribs are upon the central longitudinal line of the shell; which is not the case in *A. saccharina* or *A. stellaris.*

A. COSTATA Sowerby. Pl. 36, figs. 72–77.

Shell solid, ovate, conical, apex a little in front of the middle; slopes convex or straight; surface having 17 to 25 strong unequal ribs, the ends of which denticulate the margin.

Inside white, with a narrow gray border, often having black scallops between the marginal projections; the central area is more or less clouded with brown, usually having a distinct outline. The inner surface is sometimes stained with patches of livid purplish, light olive or brown. Length 36, breadth 29, alt. 13 mill.

South Australia; Tasmania; Chatham Is.

Lottia ? costata SOWB., Moll. of Beechey's Voyage, p. 147, t. 39, f. 2, 1839.—*Acmæa costata* TENISON-WOODS, Proc. Roy. Soc. Tasm.

for 1876, p. 50, 1877 (animal).—*P. alticostata* AD. & ANGAS, P. Z. S. 1865, p. 56, t. 2, f. 11.—*Patella costata* ANGAS, P. Z. S. 1867, p. 221.

A very variable form in sculpture and coloration. The outside is dull, usually corroded; rubbed specimens which have black scallops on the border show narrow lunate black markings in the spaces between the ribs.

Figures 72, 73 are Angas's *alticostata;* the other figures are drawn from specimens before me.

A. MARMORATA Tenison-Woods. Pl. 42, figs. 66–70.

Shell irregularly ovate, low-conical, the apex eroded, at about the anterior third. Surface dull, eroded, having 7 to 10 wide ribs, often sub-obsolete. The ribs are light, interstices dull black; edge of the shell irregular.

Inside having black stripes between the ribs, which are indicated by light stripes; central area yellowish-brown, closely dotted or marbled with black. There is a snow-white line just within the muscle-scar. Length 15, breadth 13, alt. 5 mill.

Tasmania; Port Jackson, N. S. Wales, Australia.

Acmæa marmorata TENISON-WOOD, Proc. & Rep. Roy. Soc. Tasm. for 1875, p. 156, 1876; Ibid. 1876, p. 53, 1877.

May be known at once by the peculiarly dotted, marbled interior. There is a variety which may be called *submarmorata*, in which the inside is not distinctly radiately striped around the central area, the white line within the muscle-impression is narrow or obsolete, and the outer surface has numerous rather narrow riblets in place of the few wide ribs of the typical form. This variety is from Port Jackson. It is represented on pl. 42, figs. 69, 70.

A. RUGOSA Quoy & Gaimard. Pl. 37, figs. 5, 6.

Shell ovate, conic, with radiating rugose unequal ribs; margin crenulated; broadest behind; apex sub-median, acute; inside bluish, border and center blackish-chestnut.

Length 16, breadth 12, alt. 6 mill. (*Q.*)

Amboyna.

Patelloida rugosa Q. & G., Astrol. p. 366, t. 71, f. 36, 37.

Compare *A. lacunosa* Rve.

A. LACUNOSA Reeve. Pl. 37, figs. 7, 8, 9, 10, 11.

Shell small, oblong, apex near the front third, or nearer the middle; surface dull, lusterless, rough and irregular, having 18 to 24 rude, strong, usually unequal ribs, which denticulate the margin; the ribs are whitish, interstices dull-black; sometimes the whole, shell is of a light buff-tint.

Inside white with either a blue or a pink tint; central area irregularly clouded with rust-brown or black, showing through a thin white layer; border white, black, or dotted.

Length 14, breadth 10, alt. 4½ mill.

New Zealand.

Patella lacunosa REEVE, Conch. Icon. f. 120, 1855.—*Acmæa corticata* HUTTON, Man. N. Z. Moll. p. 89, 1880.—? *Fissurella rubiginosa* HUTTON, Cat. N. Z. Moll. p. 42, *teste* Martens Zool. Rec. x, p. 150.

A small, roughly sculptured species, with much the aspect of *Siphonaria.* This likeness is increased by the frequent unsymmetrical distortions of the shell. Reeve figured a clear buff specimen; but this coloration is comparatively rare, only one out of twenty-three shells before me being wholly without the black stripes. When the outside lacks stripes, the inside is very light, its border without dark dots. The young often show dark rays within like *A. jacksoniensis.*

I have some doubt about the *A. rubiginosa.* It may prove a distinct species, being shorter and rounder. The original description is given in the MANUAL Vol. XII, p. 216. See also *tom. cit.* p. 291.

A. CINGULATA Hutton.

Shell moderately thick, white, oval, conical; apex about one-third of the length from the anterior end; about 30 to 50 low radiating ribs. Interior white, the margin light brownish-pink, banded with white. Length ·56, breadth ·45, alt, ·2 inch. (*Hutton.*)

Lyttelton and Dunedin, New Zealand.

A. cingulata HUTTON, Trans. N. Z. Institute, xvi, p. 215, t. 11, f. 5 (dentition), 1884.

The shell much resembles that of *A. lacunosa* Rve. (=*corticaria* Hutton), but the ribs are finer and the margin differently colored; the teeth, however, are very different; it may prove to be a variety of *A. rubiginosa,* the dentition of which is not yet known. (*Hutton.*)

A. CONOIDEA Quoy & Gaimard. Pl. 37, figs. 84, 85.

Shell ovate, arcuate, decidedly conical, apex obtuse, rounded; ashen; inside corneous-brown, margin spotted.

Length 13, breadth 10, alt. 11 mill. (*Q.*)

Port Lincoln, S. Australia, on stones at low water; *Sow & Pigs Reef, Port Jackson.*

Patelloida conoidea Q. & G., Astrol., p. 355, t. 71, f. 5–7.— *Acmæa conoidea* ANGAS, P. Z. S. 1865, p. 186.

A. CALAMUS Crosse & Fischer. Pl. 37, figs. 3, 4.

Shell small, round-oval, erectly conical, apex in front of the middle, slopes nearly straight. Surface sculptured with many acute, unequal radiating riblets; dull pinkish-white, often with some irregular dark brown markings.

Inside white, lilac or pink-tinted, central area more opaque white with a small brown spot in its depth; border sparsely spotted with brown. Length 10½, breadth 9, alt. 5 to 6 mill.

Port Lincoln, on stones at low water; *Tasmania.*

Patella calamus C. & F. Journ. de Conchyl. 1864, p. 348; 1865, t. 3, f. 7, 8.—*Acmæa calamus* ANGAS, P. Z. S. 1865, p. 186.

A. ALBA Tenison-Woods. Pl. 42, figs. 76, 77, 78.

Shell broad, oval, depressed, scabrous, thin, white, subshining; apex sub-median, acute; radiated with numerous small, acute, unequal scaly-granose ribs, gathered into groups; interstices delicately and most closely undulose-striate. Inside shining-white, sometimes rayed or clouded with pale-brown; no spatula. Margin acute, slightly undulating, elegantly fringed with a pale-tawny line.

Length 26, width 22, alt. 7 mill. (*T.–W.*)

North Coast of Tasmania.

A. alba T.–W. Proc. Roy. Soc. Tasm. for 1876, p. 155, 1877.

A white, silky species; porcellanous inside and delicately margined with light-brown, not unlike the Chinese umbrella-shell, but smaller. The fine scabrous ribs are gathered sometimes into a bundle, which thus forms a compound rib. It is very different from any other southern form, and rare. (*T.–W.*)

The single specimen of this species before me does not agree very well with the original description. This specimen is figured on my plate.

A. PETTERDI Tenison-Woods. *Unfigured.*

Shell broadly ovate, tumid, depressed, apex acute and submarginal; shining, dull white, very closely undulately striate with concentric growth lines and indistinctly radiated with wide, rude tawny interrupted sulci. Margin acute, elegantly fringed within with chestnut or tawny; inside white, clouded with pale chestnut; spatula tawny, sharply defined. Length 22, breadth 20, alt. 7 mill. (*T.–W.*)

Northwest coast of Tasmania.

A. petterdi T.–W., Proc. Roy. Soc. Tasm. for 1876, p. 155, 1877.

Larger than *A. septiformis*, an old enlarged specimen of which it somewhat resembles. It is dull white and shining, with the lines of growth very distinctly marked.

A. SEPTIFORMIS Quoy & Gaimard. Pl. 37 figs. 93, 94.

Shell oval, convex, very delicately radiately striated, tessellated with green or white, ornamented with radiating brown lines; inside blue or whitish, lineolate with brown.

Length 14, breadth 12, alt. 6 mill. (*Q.*)

King Georges Port, W. Australia, Kiama and near New Castle, New South Wales; Tasmania; on rocks between tides.

Patelloida septiformis Q. & G., Astrol., p. 262, t. 71, f. 43, 44.— *Tectura septiformis* ANGAS, P. Z. S. 1867, p. 220.—*Acmæa septiformis* TEN.-WOODS, Proc. Roy. Soc. Tasm. for 1876, p. 50, 1877, (animal.)

A. scabrilirata Angas, and *A. cantharus* Rve. are considered synonymous by Tenison-Woods.

A. LACINIATA Reeve. Pl. 35, figs. 36, 37.

Shell oblong, ovate, rather sharply convex, laterally slightly compressed, radiately densely elevately striated; whitish, reticulately rayed everywhere promiscuously with numerous fine red lines, network around the apex interruptedly open; interior bluish-white. (*Rve.*)

Australia.

P. laciniata RVE. Conch. Icon. f. 100, 1855.

Compare the following species.

A. CANTHARUS Reeve. Pl. 37, figs. 1, 2.

Shell ovate, rather thin, convex; apex very anterior, sharp, hooked; smooth; black, irregularly blotched with white; interior blackish-chestnut. (*Rve.*)

New Zealand; Tasmania?

P. cantharus Rv., Conch. Icon. f. 131, 1855.—*A. cantharus* HUTTON, Man. N. Z. Moll. p. 88, 1880.

A. SCABRILIRATA Angas. *Unfigured.*

Shell small, thin, subovate, a little planate ; outside of a whitish or gray color, variously maculated and penciled, sometimes streaked ; very elegantly ornamented with more or less distant, most minutely granulose, very acute radiating riblets, the interstices wide, flat. Apex curving forward, at the third or fourth of the length. Inside very shining bluish-green, varied with reddish-brown ; margin wide, tesselated or penciled ; central area rarely conspicuous. (*Angas.*)

Length 12 mill.

Port Jackson ; Port Phillip ; Port Lincoln ; Hobson's Bay, Vict. ; Holdfast Bay, St. Vincents Gulf.

A. scabrilirata ANG., P. Z. S. 1865, p. 154 ; *l. c.*, p. 186 ; *l. c.* 1867, p. 220.

This small and tender, but exceedingly beautiful species is generally more or less abraded ; but when perfect is easily recognized by the sculpture, which consists of distant, extremely slender riblets, each of which consists of, or is surmounted by, a series of minute granules. A rare variety is striped like the young of *A. pelta* (*A. strigillata* Nutt) ; but in general it is more or less, mottled, sometimes delicately penciled, like *A. fascicularis* Mke., from the Gulf of California.

May be a form of *A. septiformis.*

A. CHATHAMENSIS Pilsbry. Pl. 35, figs. 43, 44, 45, 46.

Shell oval, depressed, apex within the middle third of the length ; radiately striated ; interruptedly banded and spotted with umber on a white ground. Spatula chestnut, well-defined ; border wide, light with brown dots and lines.

The surface has even rather fine, close and obsolete radiating riblets. The coloration is peculiar, consisting of dots, spots and stripes formed of interrupted brown lines. The stripes when present are 8 or 10 in number.

The inside is white or suffused with yellow, with a brown central area and a wide border variously marked.

Length 30, breadth 24, alt. 6½ mill.

Length 24, breadth 20½, alt. 8½ mill.

Chatham Is.

This very pretty white species is quite different from any known to me from any part of the world.

A FLAMMEA Quoy & Gaimard. Pl. 37, figs. 78–83.

Shell small, ovate-conic, very delicately radiately striated, buff, flamed and reticulated with brown.

Aperture brown, white or yellowish, with a blackish border. Length 10, breadth 8, alt. 5 mill. (*Q.*)

Hobart-town, Tasmania; Isl. of Guam, Marianne Archipel.

Patelloida flammea Q. & G., Astrol. p. 354, t. 71, f. 15–24.— *Acmæa flammea* TENISON-WOODS, Proc. Roy. Soc. Tasm. for 1876, p. 51 (animal).

The elliptical, convex and turgidly conical form of the shell, as well as its greater solidity will separate this species from *A. septiformis.* Mr. Tenison-Woods refers *A. subundulata* Angas to this species, with doubt.

A. SUBUNDULATA Angas. *Unfigured.*

Shell small, thin, oval, elevated; outside of a pale brownish-corneous color, variously maculated or streaked with brown; radiating liræ obsolete, a trifle undulating; growth-striæ very close; apex scarcely curved forward, more or less anterior, at a third or two-fifths the shell's length. Inside brownish, variously maculated or streaked with blackish-brown; spatula usually dark; margin scarcely apparent. Length 13, breadth 10, alt. 5½ mill. (*Angas.*)

Port Jackson; Port Phillip; Port Lincoln; Hobson's Bay, Southern Australia; between tide marks.

A. subundulata ANG., P. Z. S. 1865, p. 155; *l.c.* p. 186; ibid. 1867, p. 220.

A variation has the inside paler, radiating streaks narrow.

A. PILEOPSIS Quoy & Gaimard. Pl. 37, figs. 90, 91, 92.

Shell ovate-convex, very finely radiately striated; blackish, dotted and netted with whitish; apex recurved to the margin; inside bluish, margin black, the center of a chestnut color. (*Q.*)

Length 18, breadth 14, alt. 8 mill.

Bay of Islands and the French Pass to Dunedin, New Zealand; Auckland Is.

Patelloida pileopsis Q. & G., Voy. Astrol., p. 359, t. 71, f. 25–27. —*Acmæa pileopsis* HUTTON, Manual N. Z. Moll., p. 88, 1880.

A. ORBICULARIS Quoy & Gaimard. Pl. 37, figs. 95–99.

Shell conical, orbicular, transversely striated, reddish-green, marked with radiating brown or reddish rays; inside bluish; vertex submedian. Length 18, alt. 6 mill. (*Q.*)

Island of Vanikoro.

Patelloida orbicularis Q. & G., Astrol. iii, p. 363, t. 71, f. 31, 32; var., f. 33, 35.

A variety (pl. 37, figs. 95–97) from Amboyna is less rounded, more oblong.

A. JACKSONIENSIS Reeve.　Pl. 42, figs. 71–75.

Shell ovate, conical, the apex near the middle or somewhat in front of it; surface smooth (or obsoletely radiately striated), dull, usually corroded or incrusted; color whitish rayed with brown.

Inside conspicuously rayed with brown and white; the central area variously mottled with brown, or continuing the rays; border narrow, scarcely different from the rest of the inside layer in color.

Length 19, breadth 15½, alt. 8 mill.

Port Jackson, Australia, on rocks at low tide.

Patella Jacksoniensis Rv., Conch. Icon. f. 127, 1855.—*Tectura jacksoniensis* ANGAS, P. Z. S. 1867, p. 220.

The smooth exterior and prominently rayed interior are the more striking characters of this species. The central area inside is variously clouded; often the rays are continued into it. When the center is entirely dark it is peculiar in shape; see fig. 75. There is often a tendency to form a white line just within the muscle-impression. A depressed specimen measures, length 18, alt. 5½ mill.

Var. MIXTA Reeve.　Pl. 35, figs. 32, 33.

Shell ovate, rather thin, conoid, compressed at the sides; apex rather anterior, obsoletely decussately striated; peculiarly mottled with black and yellow, variegated in the interior.

A thin, peculiar mottled shell, with somewhat the aspect of our northern *P. testudinalis.* (*Rve.*)

Port Phillip, Australia.

P. mixta RVE., Conch. Icon., f. 129, 1855.

Seems to be a synonym or variety of *A. jacksoniensis.*

A. CRUCIS Tenison-Woods.　Pl. 37, figs. 12, 13, 17, 18, 19.

Shell oval, conical, apex somewhat in front of the middle; slopes nearly straight; surface smooth, *without radiating sculpture*; lines of growth fine, regular. Color a dead-white, with white apex, surrounded by a small brown ring, from which *four brown stripes,* (in the direction of major and minor axes of the shell) radiate.

These stripes sometimes do not reach to the basal margin; sometimes they split, and additional stripes appear in the intervals.

Inside white, usually showing the brown stripes faintly through; central area usually brown or outlined with brown, clouded with light blue in the middle. Length 22, breadth 17, alt. 9 mill.

Tasmania.

A. crucis T.-W., Proc. Roy. Soc. Tasm. for 1876, p. 52, and animal p. 53, 1877.

Perfect, unworn specimens do not show the maltese cross of brown at the apex, it being covered by the outer layer; and they have a narrow brown border.

The figures represent specimens somewhat worn.

A. ELONGATA Quoy and Gaimard. Pl. 37, figs. 86, 87.

Shell small, ovate-elongated, fragile and pellucid; subconvex, smooth, greenish, ornamented with longitudinally reticulating reddish lines; inside white; apex marginal. Length 6, width 4 mill. (*Q.*)

King George Sound, S. W. Australia.

Patelloida elongata Q. & G., Astrol., p. 358, t. 71, f. 12–14.

A. PUNCTATA Quoy & Gaimard. Pl. 37, figs. 88, 89.

Shell small, oval, fragile, convex, smooth, whitish or buff, very delicately dotted with reddish; inside white, apex obtuse, at the margin. Length 6 mill. (*Q.*)

King George's Port, S. W. Australia.

Patelloida punctata Q. & G., Astrol., p. 365, t. 71, f. 40–42.

A. FRAGILIS Quoy & Gaimard. Pl. 37, figs. 14. 15.

Shell membranaceous, pellucid, ovate, flattened; smooth, green, ornamented with concentric brown rings; inside with an emerald ring around the muscle-impression, margin brown.

Length 15, width 12, alt. 2 mill.

The apex is anterior, sub-marginal, a little to the left of the middle.

East Coast of the North Island, New Zealand, under stones.

Patelloida fragilis Q. & G., Voy. Astrol. iii, p. 351 t. 71, f. 28–30. —*Lottia fragilis* GRAY, Dieff. N. Z. ii, p. 240,—*Acmæa fragilis* HUTTON, Man. N. Z. Moll. p. 89, 1880.—*Patella solandri* COLENSO, Tasmania Journal of Natural Science ii, pp. 226, 250, 1841; Trans. N. Z. Institute, xiv, p. 168, 1882.—*P. unguis-almæ* LESSON, Voy. de la Coquille, Zool. ii, p. 420.

This is one of the most peculiar and distinct species. In its flat, scale-like form and green color it is unlike anything else.

A. SQUAMOSA Quoy & Gaimard.　Pl. 35, figs. 34, 35.

Shell orbicular, subplane, fragile, radiately very delicately striated, painted with green and brown areoles; inside bluish, margin blackish, center chestnut colored.　Length 16, breadth 14, alt. 4 mill. (*Q.*)

Isle of France.

Patelloida squamosa Q. & G., Astrol. iii, p. 260, t. 71, f. 38, 39.

VII. SPECIES OF UNKNOWN HABITAT.

The original figures and descriptions of these are copied.　I have been able to identify none of them, and they have not been noticed by other authors.

A. ACHATES Reeve.　Pl. 35, figs. 38, 39.

Shell ovate, rather thin, convexly depressed, radiately densely striated, striæ here and there finely corded; intense black, irregularly variegated with lightning-marked white rays; interior bluish, with a broad black variegated border.　(*Rve.*)

Habitat unknown.

P. achates RVE., Conch. Icon., f. 123, 1855.

A. LENTIGINOSA Reeve.　Pl. 37, figs. 20, 21.

Shell ovate, convex, apex inclined anteriorly, obtuse, radiately striated, striæ more or less eroded; whitish, interruptedly rayed and promiscuously wave-freckled with blackish chestnut; interior bluish-white.

This species is well-characterized by the promiscuous wave-freckled style of its dark chestnut painting.　(*Rve.*)

Habitat unknown.

P. lentiginosa RV., Conch. Icon., f. 110.

It is very difficult, in the absence of *locality*, to identify limpets. The species from different parts of the world sometimes resemble each other so closely.　This should be compared with the form described by me as *A. chathamensis*, and with the more speckled forms of *A. borneensis.*

A. LIMA Reeve.　Pl. 33, figs. 86, 87.

Shell oblong-ovate, convex, apex anterior, rather hooked, radiately closely ridge-striated, striæ very minutely prickly-scaled; greenish-olive, interior bright, blue green, purple-brown at the margin.　(*Rve.*)

Habitat unknown.

A. lima Rv., Conch. Icon. f. 144, 1855.

A. NIMBUS Reeve. Pl. 35, fig. 30, 31.

Shell ovate, slightly attenuated anteriorly, rather sharply convex ; apex somewhat anterior ; everywhere decussately wave-striated, the radiating striae being the stronger ; olive, elegantly rayed with faint yellow, purplish around the apex. The rays of this species are very softly expressed, and have more the appearance of rays of light than is presented by any other species, whilst the surface striae are finely waved throughout. (*Rve.*)

Habitat unknown

P. nimbus RVE., Conch. Icon., f. 143.

A. UNCINATA Reeve. Pl. 33, figs. 91, 92.

Shell ovate, sharply conoid, apex a little hooked anteriorly ; radiately profusely finely ridged, interstices obscurely cancellated ; whitish, tessellated or diagonally streaked with blackish-brown ; interior bluish-white, transparent, tessellated and stained with faint chestnut. (*Rve.*)

Habitat unknown.

P. uncinata RVE., Conch. Icon. f. 141.

Spurious Species.

Tectura pusilla Jeffr.=COCCULINA.
Tectura adunca Jeffr.=COCCULINA.
Tectura galeola Jeffr.=COCCULINA.
Nacella peltoides Cpr. belongs to SIPHONARIIDÆ.
Tectura tahitensis and T. radiata Psc.=PATELLA.

Genus SCURRIA Gray, 1847.

Scurria GRAY, P. Z. S. 1847, p. 158 ; Guide Syst. Dist. Moll. B. M. p. 171.—DALL, Amer. Journ. Conch. vi, p. 262.—Not *Scurria* Cpr.—*Helcion, Lottia* and *Acmæa*, in part, of authors.

Shell patelliform. Animal having a branchial plume as in *Acmæa*, and an accessory branchial cordon extending entirely around the foot, or interrupted in front. Formula of radula 1 (2–1·0·1–2) 1.

This genus differs from *Acmæa* in possessing a branchial cordon, like *Patella*, in addition to the branchial plume. It agrees with *Acmæa* in dentition (*S. mesoleuca*, pl. 39, fig. 22). The subgenus *Lottia* differs from typical *Scurria* in having the branchial cordon interrupted over the head, but this distinction alone seems scarcely sufficient for generic separation.

Numerous specimens of all the following species have been examined by the author.

S. SCURRA Lesson. Pl. 39. figs. 16, 23, 24, 25, 26, 27.

Shell solid, straightly conical, elevated, the outline short oval, or nearly round, apex sub-central. Surface smooth, having fine inconspicuous radiating striæ and concentric lines of growth. Color light-brown or buff, outer layer with a waxen translucency.

Specimens are not infrequent in which the growth has been interrupted, producing an abrupt change in the color, or giving the steep, volcano-like cone, a terraced appearance. The apex, when retained, has the shape of a tiny *Lottia gigantia*. It lacks radiating striæ, but is colored with several brown stripes, as shown in figs. 23, 24 of Pl. 39. The interior is pure white.

Length 32, breadth 28, alt. 18 mill.
Length 27, breadth 22, alt. 22 mill.

12° to 41° S. Lat. West Coast of South America.

Patella scurra LESS., Voy de la Coquille, Zool., p. 421, 1830.—*Acmæa scurra* Orb. Voy. Amér. Mérid. v, p. 478, t. 64, f. 11–14.—GAY, Hist. Chile, Zool. viii, p. 252, Atlas t. 4, f. 11.—*Scurria scurra* GRAY, P. Z. S. 1847, p. 171.—DALL, Amer. Jour. Conch. vi, 263.—*Lottia pallida* SBY., Moll. Beechey's Voy. p. 147, t. 39, f. 1, 1839.—*Lottia conica* GLD., Moll. U. S. Expl. Exped. p. 346.—*Acmæa cymbula* HUPE in Gay, Historia de Chile, Zool. viii p. 252, Atlas, t. 4, f. 12, 1854.

This straightly conical species resembles the Californian *Acmæa mitra* Esch. in form. It differs in being of a buff color, in the acute, anteriorly directed apex, etc.

S. ZEBRINA Lesson. Pl. 1, figs. 10, 11.

Shell ovate, elevated, apex in front of the middle; slopes somewhat convex. Surface having 12 strong radiating ribs, about as wide as their interstices or narrower; these ribs making the margin strongly notched, when their ends are not eroded. Ground color greenish, the intervals between the ribs (or sometimes the whole surface) closely marked with triangular black blotches.

Inside smooth, muscle-scar white, scarcely impressed, the area within it chestnut colored with a darker border, becoming lighter with age, until in old shells it is almost entirely concealed by the white layer. The area outside of the muscle-scar is white or slightly clouded with brown. Edge of the shell deeply sinuated by the ribs, articulated with black ; inside of this there is a narrow blue band.

Length 54, breadth 44, alt. 20 mill.

Chili.

P. zebrina LESSON, Voy. de la Coquille, p. 417, 1830.—ORB., Voy. Amér. Mérid., p. 480, t. 65, f. 1–3.—*P. concepcionensis* LESS., *l. c.*, p. 418.—*Lottia zebrina* GLD., Moll. U. S. Expl. Exped., p. 352, t. 130, f. 460.—*Tectura zebrina* GRAY.—*Lottia variabilis* GRAY, Beechey's Voy., t. 39, f. 3, 4, but not f. 5.—*Scurria* (?) *zebrina* DALL, Amer. Journ. Conch. vi, p. 264.

The color pattern of triangular black spots is characteristic when not obscured by erosion. The eroded shell is gray, or purple tinged, with a darker apical tract.

S. PARASITICA Orbigny. Pl. 4, figs. 74, 75, 76.

Shell oval, rounded-conical, solid and strong, the apex at the anterior third, rounded off by erosion ; front slope nearly straight, posterior slope convex. Surface closely and finely radiately striated all over ; light gray or whitish, with broad and narrow radiating blackish-gray stripes, about 11 in number. Edge of shell smooth even.

Inside white, more or less clouded with chestnut inside the muscle-scar ; the yellowish-white border is rather broad, and alternately light and dark the one or the other frequently predominating.

Length 22½, breadth 17, alt. 7½ mill.

Valparaiso, etc., Chili.

P. parasitica ORB., Voy. Amér. Mérid. p. 481, t. 81, f. 1–3.— *Lottia cymbiola* GLD., Moll. U. S. Expl. Exped. p. 350, t. 29, f. 453.—Not *Patella parasitica* RVE.—*P. (Acmœa) leucophœa* PHIL., Zeitschr. f. Mal. 1846, p. 22 ; Abbild. iii, t. 2, f. 10.—? *A. punctatissima* PHIL., Zeitschr. f. Mal. 1846, p. 23 ; Abbild. iii, t. 2, f. 11.

A finely striated, arched or dome-shaped species. The basal side-margins are usually arcuate. It lives upon other shells, usually *S. zebrina.* There is but little variation in this species, and the several names are completely synonymous.

The following form seems to be closely allied, the distinction being founded mainly on the finely speckled or dotted surface—a common aspect of variation in West American limpets.

Var. PUNCTATISSIMA Philippi.　Pl. 4, figs. 71, 72, 73.

Shell ovate-elliptical, somewhat depressed, obsoletely sculptured with about 20 riblets and impressed radiating lines ; white, minutely tessellated with impressed brown dots ; apex at the front third of the length, eroded.　Inside brown, maculated with black in the cavity ; border wide, white, articulated with black ; margin very finely crenulated.　Length 14, width 11½, alt. 4 mill.

Like *leucophœa* in size and form, but differing in sculpture and coloration.　The form is elliptical, a little narrower in front.　Of ribs one sees only weak indications, and shallow furrows only toward the margin.　The rest of the outside is decorated with little brown impressed dots very regularly arranged in bands, and looking very pretty on the white ground.　There are besides inconspicuous brown rays.　The border inside is 1 to 1½ mill. wide.　(*Phil.*)

Chili.

S. MESOLEUCA Menke.　Pl. 8, figs. 96–100, 1, 2 ; Pl. 33, figs. 83, 84, 85.

Shell extremely variable in color and markings, but generally rather broad and flat, with the apex somewhat inclined anteriorly, especially in the young shell.　Outside with the apex and sometimes with a considerable portion of the shell nearly smooth ; generally with extremely fine ribs, sometimes sharp, sometimes rounded, generally slightly granulose ; sometimes with broad strong ribs ; sometimes nearly smooth with radiating lines of granules.　Sometimes intercalary ribs are found, much larger than the rest ; sometimes different plans of sculpture are seen on the same shell.　The color outside is generally olive or brownish-green ; sometimes without marking, generally with white lines either radiating or broken up ; often with white patches tesselating with the brown ; or changing from one pattern to another.　Inside, the shell is generally whitish about the middle, (whence the name) with more or less of a bluish-green tinge, sometimes dark-green, sometimes brownish, sometimes with an element of ochre-yellow more or less mottled.　There is almost always a large dark spot at the body mark, of a brownish-olive green, in which sometimes the brown, sometimes the dark-green predominates.　The body stain is irregularly and slightly gathered into points ; the head mark is generally shown by a stain shaped like a sector, bounded by two radii from the apex, about 70° apart.　The margin is generally broad, occasionally very narrow, bounded inside by a greenish line ; ordinarily tessellated with brown and white, sometimes with green or yellow ; not unfrequently with very slight markings of white, or none at all, in which case the color is either dark-greenish-brown, (*P. striata Rve.,*) or with intermediate stages to very light-greenish-white. (*Cpr.*)

Length 34, breadth 28, alt. 8 mill.

Central America to Gulf of California.

Acmæa mesoleuca Mke., Zeitschr. f. Mal. 1851, p. 38.—Cpr. Maz. Cat., p. 203.—*A. mutabilis* (part) Mke. *l. c.,* p. 37.—*P. diaphana* Rv., Conch. Icon., f. 61.—*Lottia patina?* C. B. Ad., Pan. Cat., p. 241.— *Lottia pintadina* (part) Gld., Exped. Sh., p. 9.—*P. striata* Rve., Conch. Icon. f. 99.—*P. vespertina* Rv. *l. c.* f. 67.—*Scurria mesoleuca* Dall., Amer. Journ. Conch. vi, p. 264, t. 15, f. 19 (dentition).

This variable species has much in common with *Acmæa patina ;* that species, however, is not green inside. Specimens from the Galapagos Is., are nearly typical. They have been described by Reeve as *Patella striata,* (not *P. striata* Quoy). See Pl. 8, figs. 100, 1.

The *P. vespertina* of Reeve (Pl. 8, figs. 98, 99) is doubtless synonymous.

Subgenus Lottia (Gray) Carpenter.

Lottia (Gray mss.), Sowb., Genera of Shells, pl. 42, fig. 1.— Carpenter, Journ. de Conch. 1865, p. 140.—Am. Journ. Conch. ii, p. 342, 1866.—Dall, Amer. Journ. Conch. vi, p. 260.—*Tecturella* Cpr., Smiths. Check List W. C. Sh., p. 3, 1860.—*Tecturina* Cpr., Smiths. Rep. 1860, p. 219.—*Lecania* Cpr. mss., see Amer. Journ. Conch. ii, p. 343.

The branchial cordon is interrupted over the head. Animal otherwise as in *Scurria.* Dentition, pl. 38, fig. 3.

S. GIGANTEA Gray. Plate 38.

Shell large, solid, oval, depressed, the apex near the front margin ; outer surface eroded, of a spongy texture, dull brown, gray toward the summit. Inside having a black rim around the margin, deep chestnut brown outside of the muscle-impression, which is strong, bluish or purplish-white. Central area chestnut brown, more or less mottled with white, rarely entirely white.

Length 75, breadth 55–60, alt. 17–20 mill.

San Francisco, Cal., to Panama.

5

Lottia gigantea GRAY, in SOWERBY, Genera Sh., f. 1.—REEVE, Conch. Syst., f. 1.—CARPENTER, Amer. Journ. Conch. ii, p. 343.—DALL, *l. c.* vi, p. 260, t. 15, f. 20 (animal and dentition).—*Acmæa scutum* AUCT., not of Esch. nor d'Orb.—*Tecturella grandis* Gray, CPR., Smiths. Chk. List no. 176; Brit. Asso. Rep. 1861, p. 137.—*Patella kochi* PHILIPPI, Abbild. u. Beschreib. iii, Patella, t. 1, f. 1, Jan., 1849.

Young specimens (fig. 4) have fine, nearly obsolete radiating striæ and are roughened by low, obliquely radiately riblets in front and at the sides, which on the posterior half of the shell become broken into low rounded tubercles or bosses arranged in curved rows obliquely descending from the central line of the back.

This is the largest and handsomest of the Californian limpets. Philippi's excellent illustration of it seems to have been unknown to writers on west coast shells.

Family *LEPETIDÆ* Gray.

Lepetidæ GRAY, Guide Syst. Dist. Moll. B. M., p. 172, 1857.—*Patellidæ* and *Tecturidæ*, in part, of authors.

Shell conical, patelliform, with subcentral or anterior apex; surface feebly sculptured, the edge smooth; muscle-impression as in *Patella*. Embryonic shell spiral.

Animal without external branchiæ; radula provided with a rhomboidal cuspidate rhachidian tooth; no lateral teeth; uncini slender, two each side, their cusps simple or fringed. Formula 2·0·1·0·2.

It will be noted from the formula given that the side-teeth are considered *uncini* rather than true *laterals*. The lateral teeth are either aborted or represented by the lateral cusps of the rhachidian tooth, which is, if this be the case, a compound body formed by coalescence, as is the case in *Phasianella* (*Orthomesus*) *virgo* Ang. I am not inclined, however, to believe this to be the case in *Lepeta*.

The very different dentition of the genus *Lepetella* renders its reference to this family somewhat doubtful.

Synopsis of genera and subgenera.

Subfamily LEPETINÆ Dall.

Radula having a large central tooth with several cusps, and two side-teeth on each side.

Genus LEPETA Gray, 1847.

Shell patelliform, the embryonic nucleus spiral, lost in the adult; apex in front of the middle; no internal septum. Animal without external branchiæ; having the muzzle produced into a labial process on each side. Dental formula 2·0·1·0·2, the uncini narrow, erect, provided with cusps. Type, *L. cæca.*

Section *Lepeta* s. str. Apex erect; anterior terminations of the great muscle-scar in front of the apex. Color whitish or light-brown. Surface granulate. Apex of rhachidian tooth 5-cuspidate, middle cusp large, lateral cusps small; uncini subspatulate at apex, obtuse, not ciliated.

Section *Cryptobranchia* Midd. Apex inclined forward, the anterior terminations of the muscle-scar not in front of it. Surface not granulate; color whitish. Apex of rhachidian tooth tricuspidate, the cusps nearly equal; uncini spatulate. Type, *C. concentrica.*

Subgenus PILIDIUM Forbes.

Apex of shell anterior; surface delicately radiately ribbed and granulate; color dark orange or reddish.

Apex of rhachidian tooth tricuspidate, the middle cusp much the largest, side cusps small, triangular, disjoined from the central one at their bases; uncini elongated, apex lanceolate, its inner margin densely ciliate.

Genus PROPILIDIUM Forbes & Hanley, 1849.

Shell patelliform, elevated; the spiral nucleus retained in the adult; apex central; surface cancellated; inside having a small triangular plate situated deep in the cavity, as in *Puncturella.* Type, *P. ancyloide.*

Rhachidian tooth tricuspidate, the central cusp rather long, side cusps small, separated from the middle cusp at their bases. Uncini with the cusps finely denticulated.

Subfamily LEPETELLINÆ Dall.

Radula having three teeth on each side of the simple central tooth.

Genus LEPETELLA Verrill, 1880.

Shell patelliform, small, smooth, oval, conical, with a subcentral apex, spiral in the young. Animal having distinct eyes. Radula having lateral teeth and uncini, formula 2·1·1·1·2. Type, *L. tubicola.*

Subfamily LEPETINÆ Dall.

Genus LEPETA Gray, 1847.

Lepeta GRAY, P. Z. S. 1847, p. 168.—DALL, Amer. Journ. Conch. v, p. 140.

Dentition of *L. cæca*, pl. 40, figs. 31, 31 ; enlarged apex showing spiral embryonic shell, pl. 40, fig. 32.

L. CÆCA Müller. Pl. 40, figs. 29–32.

Shell rather straightly conical, apex erect ; front slope one-half the length of shell ; surface sculptured with fine close radiating striæ, rendered granulous by the intersection of equally close low thread-like concentric striæ ; color light-brown, sometimes having a pinkish tint.

The outline is oval ; front slope nearly straight, posterior slope a little convex, both often slightly concave above. The beaded sculpture is most developed toward the upper part of the cone. Inside dirty white or pink tinged. Apex generally eroded ; basal margins of shell level. Length 14, breadth 11, alt. 5 mill.

Arctic Ocean, and North Atlantic, South to Massachusetts Bay, Scotland and Denmark ; Off Sea Horse Is., near Point Barrow, Alaska, and North from Bering Strait.

Patella cæca MULL., Prodr. Zool. Dan., p. 237.—*Lepeta cæca* GRAY, P. Z. S. 1847, p. 168.—JEFFREYS, Brit. Conch. iii, p. 252.— DALL, Amer. Journ. Conch. v, p. 141 ; Proc. U. S. Nat. Mus. i, p. 334, 1878.—GOULD, Invert. of Mass. 2d edit., p. 270, f. 531.— BERGH, Verh. z.-b. Ver. Wien, xxi, p. 1300.—SARS, Moll. Arct. Norv., p. 123, t. 20, f. 17.—*Patella cerea* MOLL., Grœnl., p. 16.— *Patella candida* COUTHOUY, Bost. Journ. N. H. ii, p. 86, t. 3, f. 17, 1838.—GOULD, Inv. of Mass., p. 152, 1841.—*Pilidium candidum*, STIMP., Sh. of New Engl., p. 29.—? *Lepeta franklini* GRAY, Guide Syst. Dist., p. 172.

This is a small whitish species, seen under a lens to be very distinctly and beautifully granulated.

Section *Cryptobranchia* Middendorff, 1851.

Cryptobranchia MIDD., Sib. Reise, p. 183.—DALL, Amer. Journ. Conch. v, p. 143, 1869, Proc. U. S. Nat. Mus. i, p. 334, 1878, and *l. c.* iv, p. 412.

Differs from *Lepeta s.* str. in having the apex of the shell more anterior, the surface not granulate, the median cusp of the rhachidian tooth not longer than the side cusps. Dentition of *L. concentrica*, pl. 40, fig. 35; of *L. alba*, pl. 40, fig. 40.

L. CONCENTRICA Middendorff. Pl. 40, figs. 33, 34, 35, 36, 37.

Shell depressed conical, apex directed forward; front slope one-third the length of the shell or a little less; surface faintly radiately striate (more distinctly so in young specimens), not decussated or granulose; light-brownish or greenish tinted.

The outline is ovate, a little narrower in front; front slope slightly concave, posterior slope convex. The fine thread-like radiating striæ are larger on the longer slope of the shell; they are not interrupted by concentric growth-lines, the latter being inconspicuous, or sometimes strongly impressed at intervals. Epidermis very thin, yellowish-brown, deciduous. Inside polished, white, the anterior terminations of the muscle-scar a little behind the apex. Edges of shell level, narrowly bordered with gray, especially in the young. Length 20½, breadth 16, alt. 6 mill.

North Japan, along the Aleutian Is. and along the southern coast of Alaska, southward to Puget Sound.

Patella (Cryptobranchia) cæca var. concentrica MIDD., Siber. Reise, p. 183, t. 16, f. 6, 1851.—*Cryptobranchia concentrica* DALL, Amer. Journ. Conch. v, p. 143, t. 15, f. 2, 1869; Proc. U. S. Nat. Mus. i, p. 334, 1878.—*Lepeta cæcoides? n. sp.,* CPR., Suppl. Rep. Br. Asso. Adv. Sci. 1863, pp. 603, 651.—*Lepeta cæcoides* COOPER, List Cal. Moll., p. 24.—CPR., Proc. Acad. N. S. Phila. Apr., 1865, p. 60.

Differs from *L. cæca* in the simply striated, not granulose surface, more anterior apex, larger size and more depressed form.

Var. INSTABILIS Dall. Pl. 40, figs. 44, 45, 46.

Shell depressed, apex anterior, length of front slope contained three and one-half times in the length of the shell; surface smooth, with occasional rings caused by more impressed growth-lines. Basal margin curved upward at each end. Color whitish.

The outline is shortly ovate, front and posterior slopes nearly straight, the young may be lightly striate. Shell thick, solid, muscle-impression deep. Length 14, breadth 12, alt. 4 mill.

Sitka, Alaska, in 10 fms.

? Cryptobranchia instabilis nom. prov., DALL, Amer. Journ Conch. v, p. 145, t. 15, fig. between 3b and 5b, 1869.—*C. concentrica v. instabilis* DALL, Proc. U. S. Nat. Mus. i, p. 335, 1878.

May be known by its rounded form, the ends turned upward, much as in *Clypidella pustula*. The soft parts are unknown.

There is an error in the numbers of the plate referred to by Dall. The figure between figs. 3b and 5b represents this species.

L. ALBA Dall. Pl. 40, figs. 38, 39, 40.

Shell pure white, smooth or with extremely faint striæ; solid; interior pure white; apex directed anteriorly, inconspicuous; shell arcuate before and behind.

Length of adult 24, width 17½, alt. 10 mill.

This species differs from the last in its smooth shell, greater size, pure whiteness, greater lateral compression, and generally more rounded back, from the less prominent apex. The tentacula in a specimen twice the size of a *concentrica* were not half as large. The teeth differ in the shape of the central tooth and the greatly broader cusps of the laterals, and their striation, resembling those of *Pilidium fulvum*.

Aleutian Islands; Seniavine Strait; Plover Bay, E. Siberia.

Cryptobranchia alba DALL, Amer. Journ. Conch. 1869, p. 145, t. 15, f. 3a-d; Proc. U. S. Nat. Mus. i, p. 335, 1878.—*Patella alba* AURIVILLIUS, Vega-expeditionens Vetenskapliga Iakttagelser, iv, p. 318, t. 12, f. 10, 11, 1887.

Subgenus PILIDIUM Forbes, 1849.

Pilidium FORBES, Athenæum, Oct. 6, 1849, p. 1018.—FORBES & HANLEY, Hist. Brit. Moll. ii, p. 440, 1849.—DALL, Amer. Journ. Conch. v, p. 146, 1869.—*Iothia* GRAY, (not Forbes) Guide Syst. dist. Moll. B. M., p. 172.—*Scutellina* CHENU, in part, and of SARS, Moll. Reg. Arct. Norv., p. 122, not *Scutellina* of Gray and authors.— Not *Pilidium* MIDD.,= *Capulacmæa* SARS, (see *Capulidæ.*)

The dentition of *P. fulvum* is figured on pl. 40, fig. 43.

L. FULVA Müller. Pl. 40, figs. 41, 42, 43.

Shell small, apex strongly inclined forward, near to the front margin; front slope steep, somewhat concave, posterior slope long, convex; sculptured with delicate, closely granular radiating threads. Color orange or reddish.

The form is oval; apex pointed, prominent. The riblets of the surface are delicate, thread-like, separated by intervals wider than themselves; they are closely, finely granulose.   Interior polished, orange reddish, rarely white.

Length 4·7, breadth 3·3, alt. 2 to 2·2 mill.

Coast of Scotland and Ireland ; Scandanavia, 5 to 100 fms.

Patella fulva MULLER, Prodr. Faun. Dan., p. 227.—*Pilidium fulvum* FORBES, Athenæum, 1849, p. 1018.—FORBES & HANLEY, Hist. Brit. Moll. ii, p. 441, t. 62, f. 6, 7 ; pl. AA, f. 3.—DALL, Amer. Journ. Conch. v, p. 147, 1869, t. 15, f. 4, 4a ; Proc. U. S. Nat. Mus. i, p. 335.—*Tectura fulva* JEFFR., Brit. Conch. iii, p. 250.—*Scutellina fulva* SARS, Moll. Reg. Arct. Norv., p. 122, t. II, f. 12 (dentition).—*Patella forbesii* J. SMITH, Mem. Werner Soc. viii, p. 107, t. 2, f. 3.—BROWN, Ill. Conch. Gt. Br., t. 57, f. 3, 4.—*Iothia fulva* GRAY, figs. Moll. An., p. 93, 1854.

May be known by its small reddish or dark colored shell, anterior apex and fine beaded radiating riblets.   *Patella rubella* Fab. has has been referred here as a synonym, but it is a species of *Acmæa.*

L. COPPINGERI E. A. Smith.   Pl. 39, figs. 20, 21.

Shell cap shaped, thin, sculptured with numerous fine, thread-like, granulous liræ radiating from the apex to the margin, and with fine concentric lines of growth.   The color is dirty white, varied with two or three bands of a pale slate-color which encircle the shell at irregular intervals, and are interrupted by the radiating liræ, which are white.   This feature is more apparent within the shell, where the surface is very smooth and shining.   Margin nearly simple, very crenulated by the extremities of the ridges, roundly ovate in form. Apex rather acute, not greatly curved down, and very near the anterior end.   Length 5½ mill., diam. 4½, height 2½.   (*Smith.*)

Sandy Point, Eastern part of the Sts. of Magellan, Patagonia, 9–10 fms.

Tectura (Pilidium) coppingeri SMITH, P. Z. S. 1881, p. 35, t. 4, f. 12, 12a.

This is the southern representative of the northern *Tectura (Pilidium) fulva* of Müller.   It is rather more circular than the latter ; and the color of the single specimen at hand is different. (*Smith.*)

L. EMARGINULOIDES Philippi.   *Unfigured.*

Shell minute, elliptical, rather depressed, white, thin ; vertex nearly marginal ; having about 30–36 scaly ribs ; margin subdentate.

Length 1⅔, breadth 1⅓, lines.   (*Ph.*)

Magellan.

Patella? emarginuloides PHIL., Mal. Bl. xv, p. 224, 1868.

A single specimen seen. Has the aspect of *Emarginula*, but no trace of any incision in the shell. The riblets in front are smaller, filiform; wider and more distant behind. The animal soaked in water showed elongated tentacles with eyes at their bases, but branchiæ could not be made out. (*Phil.*)

This may be referred to *Lepeta* provisionally.

Genus PROPILIDIUM Forbes & Hanley, 1849.

Propilidium F. & H., Hist. Brit. Moll. ii, p. 443, 444.—SARS, Moll. Reg. Arct. Norv., p. 123.—FISCHER, Manuel, p. 863.—DALL, Blake Report, p. 412.—*Rostrisepta* SEGUENZA.

The adult shell of this genus retains the distinctly spiral nucleus; and in the cavity of the apex it is furnished with a small plate or septum, like that of *Puncturella*.

The dentition (pl. 39, figs. 13, 14) is very similar to that of *Pilidium*. It may well be doubted whether there are any gills, although Forbes' original figure shows a pair of small plumes. This point calls for additional observation. At all events, the position of the genus is doubtless in *Lepetidæ*, for the dentition agrees with no other family of limpets.

P. ANCYLOIDE Forbes.   Pl. 39, figs. 8, 9, 10, 11, 12, 13, 14, 15.

Shell having an oval outline, compressed at the sides, rather thin, semitransparent, glossy at the apex, but elsewhere of a dull hue; sculpture, very numerous and close-set fine striæ, which radiate from the beak and are exquisitely granulated in consequence of their being intersected or decussated by equal sized concentric striæ; color dirty white, occasionally diversified by a few clear longitudinal rays or lines; beak smooth and highly polished, styliform and slender, pinched up into a minute spire of between one and two whorls, which curls downwards at the posterior end; mouth oval; of nearly the same breadth throughout; margin thin and even, minutely tuberculated in immature specimens; inside nacreous, furnished in the centre with a thin laminar partition, like the half deck of a vessel, which has its opening towards the head or anterior part; pallial scar broad.

Length 0·15, breadth 0·115 inch. (*Jeffr.*)

Coasts of Ireland and Scotland; Scandanavia; Naples, and Trapani, Sicily, 10–145 fms. Fossil in Sicilian Pliocene.

Patella? ancyloides FORBES, Ann. Mag. N. H. v, p. 108, t. 2, f. 16.—*Propilidium ancyloide* F. & H., Hist. Brit. Moll. ii, p. 443, t. 62, f. 3, 5; t. AA. f. 4.—JEFFREYS, Brit. Conch. iii, p. 254.—SARS, Moll. Reg. Arct. Norv., p. 123, t. 20, f. 18a–e.—DALL, Blake Report, p. 412, t. 31, f. 2 (dentition).—*P. ancyloides* JEFFR., P. Z. S. 1882, p. 673.—*Rostrisepta parva* SEG., *teste* JEFFR.

The small internal transverse septum is a peculiar feature of this species.

P. SCABROSUM Jeffreys. Pl. 40, figs. 47, 48.

Shell roundish-oval, expanded, rather thin, semitransparent and of a dull hue; sculpture, numerous but not close-set, slight striæ which radiate from the beak and are more or less covered with short tubercles, especially behind; there are also several concentric ridges as in the last-named species; color whitish; beak small, pinched up, incurved, and forming a minute spire of two whorls; mouth roundish oval; margin thin; inside glossy; septum thick and strong. Length 0·15, breadth 0·15 inch. (*Jeffr.*)

Adventure Bank, Mediterranean.

P. scabrosum JEFFR., P. Z. S. 1882, p. 674, t. 50, f. 6.

Differs from *P. ancyloides* in being round instead of oval, and in having much fewer and tuberculated striæ; but I am not quite satisfied that it is more than a curious variety. It somewhat resembles the young of *Gadinia garnoti;* but that shell has not the internal septum which is characteristic of the present genus.

P. AQUITANENSE Locard. *Unfigured.*

Shell very small, patelliform, conic a little elevated; thin, rather solid, opaque, a little rugose, ornamented with longitudinal striæ which are very obsolete, visible only toward the base. Basal margin continuous, smooth, irregularly level, visibly turned up at the two extremities, descending at the median part; aperture very broadly elliptical, a little contracted behind, well rounded in front. Summit subcentral, a little anterior, little elevated, recurved toward the anterior region. Length 2, width 1¾, alt. 1¾ mill. (*Locard.*)

France.

P. aquitanense LOCARD, Proc. de Malac. Francaise, Obs. sur la Faune Marine des Cotes de Fr., in Annales de la Soc. Linn. de Lyon, xxxii, 1885, p. 244 (1886.)

This form has not been figured, and no locality is given by Locard in his worthless publication. It is said to be near *P. scabrosum*

Jeffr. but less regularly elliptical, more lengthened, narrower behind ; the altitude less in proportion to the greatest diameter, the apex a little more anterior, surface less ornamented.

P. PERTENUE Jeffreys. Pl. 40, fig. 49.

Shell oval, convex, very thin and delicate, transparent, and glossy ; sculpture, none ; color whitish ; beak small, cylindrical, and incurved, forming a minute spire of two whorls : mouth oval ; margin even ; inside glossy ; septum small.

Length 0·1, breadth 0·075 inch. (*Jeffr.*)

> Off Rinaldo's Chair and Palermo, Mediterranean 162½ fms.
> Off Rhode Island, 640 fms.

P. *pertenue* JEFFR., P. Z. S. 1882, p. 674, t. 50, f. 7.—VERRILL, Trans. Conn. Acad. vi, pp. 262, 271.—DALL, Blake Gastrop., p. 412.

The young shells of *P. ancyloides* are much smaller than the species now described, are more expanded or depressed, and have the same sculpture as the adult ; they are also proportionally solid as well as of a dull hue.

The inner layers of most of the specimens are permeated by a microscopic and branching spore-like organism, perhaps of a fungoid nature.

An imperfect specimen of another small and apparently distinct species occurred also in Station 17. It has the characteristic septum, but otherwise resembles a *Lepetella*. The beak is very much shorter than in *P. pertenue ;* and the spire has barely one turn.

I have originally given the species above described the *ms.* name *tenue*. (*Jeffr.*)

The identity of the specimens collected off Rhode Island, by Prof. Verrill, is not certain.

P. COMPRESSUM Jeffreys. Pl. 40, fig. 50.

Shell differs from *P. pertenue* in being oblong intead of oval, and in being laterally compressed like *Patella (Lepetella) latero-compressa* of Rayneval, a Monte Mario fossil, and, according to Dr. Tiberi, living in the Bay of Naples ; and it is also not quite smooth, but is marked by a few slight longitudinal striæ ; the beak is proportionally longer, somewhat twisted to one side, and nearly overhangs the hinder margin, instead of being placed (as in *P. pertenue*) at about one-third of the distance from it.

Length 0·1, breadth 0·065 inch. (*Jeffr.*)

> North Atlantic.

P. compressum JEFFR., P. Z. S. 1882, p. 674, t. 50, f. 8.

P. ELEGANS Verrill. *Unfigured.*

Shell small, very thin and fragile, translucent bluish-white, rather depressed, elongated-elliptical with the recurved apex situated at about the posterior third. The nuclear whorl is very minute, smooth glassy, compressed, strongly involute and turned a little to the left, forming a complete whorl, visible in a side view. The whole surface, under the microscope, has the appearance of a very fine shagreen. This is produced by very minute, short, wavy, raised lines, which are mostly arranged in zigzag or in herring-bone style; in some parts the two sets of lines, running obliquely, cross each other at nearly right angles; on other portions one or both sets are replaced by minute punctations, or granulations. This sculpture is visible only under a strong lens or with the compound microscope. The internal lamina or septum is narrow, crescent-shaped, situated behind and some little distance below the extreme apex, and not forming an elongated channel; it is distinctly visible from the outside, owing to the translucency of the shell. (*Verrill.*)

Length 3·5, breadth 2·5, alt. 1 mill.

Off Chesapeake Bay, 1395 fms.

P. elegans VERRILL, Trans. Conn. Acad. vi, p. 205.

The animal has a short, broad ovate foot, subtruncate in front, with the edge frilled. Frontal disk rather large, broad semicircular or crescent-shaped, with the angles extending back in a large obtuse lobe on each side. Buccal area semicircular; mouth surrounded with four convex elevations, one before and one behind it, and one on each side. Tentacles slender, tapering, acute. Eyes apparently wanting. No cirri on mantle. (*Verrill.*)

Subfamily LEPETELLINÆ Dall.

Genus LEPETELLA Verrill, 1880.

Lepetella VERRILL, Amer. Journ. of Science, 3d Ser., xx, p. 396, Nov., 1880; Proc. U. S. Nat. Mus. iii, p. 375, Jan., 1881.—DALL, Proc. U. S. Nat. Mus. iv, p. 408, 1882; Blake Gastrop., p. 413.

The animal has eyes. There are seven rows of teeth, the dental formula being 2·1·1·1·2 (pl. 39, fig. 17). Soft parts otherwise as in *Lepeta*. The embryonic shell is spiral (pl. 39, fig. 19.)

L. TUBICOLA Verrill & Smith. Pl. 39, figs. 17, 18, 19.

Shell thin, white, smooth, conical with the apex acute and nearly central; aperture broad elliptical, oblong or subcircular, usually more or less warped, owing to its habitat; edge thin and simple. Sculpture none, lines of growth slight, outer surface dull white; inner surface smooth, with the pallial markings faint.

Length 3·75, breadth 3, alt. 2 mill. (*V.*)

Off Martha's Vineyard; Gulf of Mexico between the delta of the Mississippi and Cedar Keys, Fla., 130–388 fms.

L. tubicola V. & S. Amer. Jn. Sci. 1880, p. 396; Proc. U. S. Nat. Mus. 1881, p. 375; Trans. Conn. Acad. v, p. 534, t. 58, f. 29, 29a.— DALL, Proc. U. S. Nat. Mus. iv, p. 408; Blake Gastrop., p. 413, t. 25, f. 6 (Dentition.)

Young specimens show that the nucleus is subspiral, as in other *Lepetidæ.*

Family *PATELLIDÆ.*

Docoglossate gasterpods having a simply conical shell, non-spiral even in the embryo. Breathing by a cordon of branchial leaflets attached to the mantle between its thickened edge and the sides of the foot; having no cervical gill-plume. Radula having three uncini and three laterals on each side, the rhachidian tooth being either present, rudimentary or wanting; jaw developed.

The *Patellidæ* differ markedly from *Acmæidæ* and *Lepetidæ* in the gills, which form a complete or interrupted cordon, not accompanied by a cervical branchial plume, and not homologous with the gills developed in other *Prosobranchiata.*

The shells may generally be distinguished from those of the *Acmæidæ* and *Lepetidæ* by their texture and the lack of a defined internal border; but the distinction is difficult or impossible to express in words, and must be learned by actual familiarity with the objects themselves.

In the arrangement of the species and groups I have made use of the character of the *texture of the interior,* heretofore neglected by systematists, but undoubtedly of equal importance in many cases for the discrimination of groups with the character of the gill-cordon and the presence or absence of a rhachidian tooth.

The rhachidian tooth is now proven to be decidedly variable in closely allied species. It is well-developed and bears a cusp in *Ancistromesus* and many species of *Scutellastra;* is represented by a linear rudiment in *Patina, Nacella, Patinella* and *Helcioniscus.* It is apparently wanting in *Patella* s. str.

An original formula is herein used to express the arrangement of teeth upon the radula, not from any preference for novelty, but because the new method is believed to be more graphic.

The gill-cordon is probably always interrupted by a narrow hiatus at the front left side but practically it is considered "interrupted" only when absent above the neck for a considerable distance.

An excellent paper by R. J. Harvey Gibson on the anatomy of *Patella vulgata*, in Trans. Roy. Soc. Edinb. xxxii, pt. 2, p. 601–638, 5 plates, gives the most complete account yet published of the anatomy, histology and physiology of *Patella*. Illustrations of the dentition of various *Patellidæ* have been published by Dall, Sars, Hogg, Hutton and others.

Numerous classifications have been proposed for the limpets. The systems of the earlier authors, as well as of H. & A. Adams, are very crude, being founded upon the shell alone. J. E. Gray (Guide) offered a somewhat better but still very imperfect arrangement. Dr. W. H. Dall in 1871, proposed a system based mainly upon the gill-cordon and dentition. This was somewhat modified by him a decade later in Proc. U. S. Nat. Mus. iv, 1881, p. 412, 413 (1882). His modified arrangement is as follows:

A. Branchial cordon complete.

   a. With rhachidian tooth; branchial lamellæ arborescent, produced; sides of foot smooth. ANCISTROMESUS.

      Ancistromesus Dall. $_3(_1{}^{212}{}_1)_3$.

   b. Without rhachidian tooth; branchial lamellæ short, linguiform. PATELLA.

      Patella Linné. Foot smooth; branchial lamellæ subequal all around. $_3(_1{}^{202}{}_1)_3$.

      Patinella Dall. Foot with scalloped frill interrupted only in front; gills as in *Patella*. $_3(_2{}^{101}{}_2)_3$.

      Nacella Schum. Foot frilled; gills very small in front; shell peculiar; lateral teeth all bidentate. $_3(_2{}^{101}{}_2)_3$.

B. Branchial cordon interrupted in front.

   a. Sides of foot smooth. HELCION.

      Helcion Montf. Third laterals posterior, bidentate. $_3(_1{}^{202}{}_1)_3$.

      Helcioniscus Dall. First laterals anterior. $_3(_2{}^{101}{}_2)_3$.

      Patina Gray. Third laterals posterior, denticulate; shell peculiar. $_3(_1{}^{202}{}_1)_3$.

* * *

      Metoptoma Phillips (fossil.)

Dr. Paul Fischer's classification (Manuel de Conch., p. 866, 1885) is as follows:

Genus PATELLA.

> Subgenus *Patella* s. s.   Brachial cordon complete; no tubercles on the epipodial line; dentition $_3(_1{}^{202}{}_1)_3$.   *P. vulgata*, etc.
>> Section *Ancistromesus* Dall.   *P. mexicana.*
>> Section *Olana* Ads.   *P. cochlear.*
>> Section *Scutellastra* Ads.   *P. pentagona.*
>> Section *Cymbula* Ads.   *P. compressa.*
>> Section *Patellastra* Monts.   *P. lusitanica.*
>
> Subgenus *Patinella* Dall.   Branchial cordon complete; epipodial line scalloped; no central tooth; dentition $_3(_2{}^{101}{}_2)_3$.   *P. magellanica.*
>
> Subgenus *Nacella* Schum.   Animal as in the last.   Shell oval, thin, pellucid, summit anterior, submarginal.   *N. mytilina.*
>> Section? *Cellana* Ad.
>
> Subgenus *Helcion* Montf.   Branchial cordon interrupted in front.   Epipodial line papillose; dentition $_3(_1{}^{202}{}_1)_3$.   *H. pectinatus.*   Synonym *Patina* (Leach) Gray.   *P. pellucida.*
>
> Subgenus *Helcioniscus* Dall.   Branchial cordon interrupted; sides of the foot smooth; dentition $_3(_2{}^{101}{}_2)_3$.   *H. variegatus.*

Genus TRYBLIDIUM Lindstr., 1880.   Shell like *Patella;* muscle-scar broken into a number of separate impressions (fossil.)

> Subgenus *Palæacmæa* Hall, 1873.   Shell like *Scurria;* muscle-scar like *Tryblidium* (fossil.)

Dr. Fischer places *Metoptoma* in *Capulidæ* on account of the posterior apex, which is unlike all docoglossate limpets.

It is evident that a great mass of material must be examined before a just appreciation of the constancy of the characters used to separate groups in this family can be attained.

A survey of all available sources of information upon the soft parts and radulæ has convinced me, against my preconceived ideas, that the presence or absence of a rhachidian tooth and the continuity or interruption of the branchial cordon are not sufficiently constant to be used as characters for the separation of genera.   In some cases it is evident that they are scarcely specific.   The radulæ should be thoroughly re-examined, as many of the published figures are not sufficiently accurate to be of much use.

Synopsis of Groups of Patellidæ.

A. Two inner lateral teeth on each side anterior.

Subgenus PATELLA Linné, 1758 (restricted.)

Branchial cordon complete; sides of foot having no epipodial projections. Two inner lateral teeth on each side anterior, the rhachidian tooth present or absent. Apex of the shell near the center. Type *P. vulgata*.

> Section PATELLA *s. str.* Inner layer of the shell subtranslucent, exhibiting when closely examined a concentrically fibrous texture; more or less iridescent. Radula without a rhachidian tooth; formula $_3(_1{}^{202}{}_1)_3$. Type *P. vulgata*. *Cymbula* Ads. and *Patellastra* Monts. are synonyms.

> Section SCUTELLASTRA Ads., 1858. Inner layer of the shell opaque, porcellanous, not iridescent. Radula either with or without a rhachidian tooth; formula $_3(_1{}^{202}{}_1)_3$ or $_3(_1{}^{212}{}_1)_3$. *Olana* Ads. is a synonym.

> Section ANCISTROMESUS Dall, 1871. Inner layer of the large, massive shell porcellanous, opaque. Rhachidian tooth of the radula developed, bearing a cusp; formula $_3(_1{}^{212}{}_1)_3$. Type *P. mexicana*.

Subgenus HELCION Montfort, 1810.

Branchial cordon interrupted in front; side of foot smooth, without epipodial processes. Inner two lateral teeth on each side anterior, no rhachidian tooth. Shell oval, apex anterior. Formula of teeth $_3(_1{}^{202}{}_1)_3$.

> Section HELCION *s. str.* Shell oval, with anterior apex; surface sculptured with scaly radiating ribs. Type *H. pectinatus*.

> Section PATINA (Leach) Gray, 1840. Shell with anterior or subcentral apex; radiately striated, polished.

B. One inner lateral tooth on each side anterior.

Subgenus NACELLA Schumacher, 1817.

Branchial cordon complete; sides of foot bearing a scalloped epipodial ridge. One inner lateral tooth on each side anterior; rhachidian tooth none or rudimentary. Interior of shell having a satin-like or metallic luster.

Section NACELLA s. str. Gills very small in front. Shell oblong, thin, the apex curved forward, near or at the anterior extremity of the shell. Type *P. mytilina.*

Section PATINELLA Dall, 1871. Gills equally developed all around. Shell solid, colored, ribbed, the apex subcentral or anterior. Type *P. magellanica.*

Subgenus HELCIONISCUS Dall, 1871.

Branchial cordon interrupted; sides of the foot smooth, lacking epipodial processes. One inner lateral tooth on each side, anterior. Shell solid, having the apex subcentral or subanterior, inner layer subtranslucent, more or less iridescent or satiny. Type *H. variegatus* Rve.

Resembles *Patinella* in texture of shell and dentition, but lacks an epipodial ridge, in the latter respect resembling *Patella.* It differs from both in having the branchial cordon interrupted in front.

Genus PATELLA L., 1758.

Patella L., p. Syst. Nat. x, p. 780 (in part).—*Eruca* TOURNEFORT. —*Patellites* WALCH.—*Patellaria* LLHWYD.—? *Goniclis* RAF. *olim.* —*Patellus* MONTFORT.

Subgenus PATELLA (restricted.)

The subgenus is here considered to=*Patella*+*Ancistromesus* of Dr. Dalls's arrangement. It is one of the best-defined groups of the family, being characterized by (1) the continuous branchial cordon, (2) smooth sides of the foot, having no epipodial ridge or processes, and (3) having a peculiarity of the radula found in none of the other groups except *Helcion*+*Patina;* viz., the two inner lateral teeth on each side are unicuspid and situated *in front of* the third laterals, which are larger and have several (generally three) cusps. This disposition is easily understood by reference to pl. 52, fig. 1 representing the odontophore of *Patella vulgata.* In some species a central or rhachidian tooth is developed, and when this is the case it is placed on the same level with the inner laterals. See under section *Ancistromesus.*

A reference to the synopsis of groups on page 79 shows that the subgenus consists of three sections, of which the first is

Section Patella (restricted.)

Cymbula H. & A. Adams, Gen. Rec. Moll., p. 466, is a synonym, its type being *P. compressa* L.

Rhachidian tooth of the radula absent.

Inner layer of the shell subtranslucent, exhibiting when closely examined a *fibrous texture;* usually more or less iridescent. Distribution: European seas, West Africa and the adjacent islands.

The oceanic and west African allies and varieties of *P. cœrulea* are much in need of revision. Too many species have been made, most of which are here retained for want of sufficient material to show their actual specific affinities.

P. FERRUGINEA Gmelin.   Pl. 53, figs. 1, 2, 3 ;   pl. 17, figs. 23, 24.

Shell oval, conical, solid, the apex in front of the middle ; slopes straight or convex ; *roughly sculptured with numerous* (44–50) *strong, unequal rounded radiating ribs*, which are wider than their interspaces, and which strongly denticulate the margin.   Dull and lusterless, ashen, more or less stained with brown.

Interior bluish-white, porcellanous ; muscle-scar deeply impressed ; central area thick, callous, opaque-white, its border well-defined. Margin strongly fluted, having a brown line at the edge.

Length 62, breadth 50, alt. 20 mill.

Mediterranean Sea, from the Ægean to Spain.

P. ferruginea GMEL., Syst. p. 3706.—WEINKAUFF, Conchyl. des Mittelm. ii, p. 401.—*P. lamarckii* PAYR., Moll. de Corse p. 90, t. 4, f. 3, 4.—DESH., Exped. Sci. de Morée iii, p. 133.—*P. plicata* COSTA, Catal. Sist. p. 119.—*P. costosoplicata* REEVE, Conch. Icon. f. 14.— HIDALGO, Journ. de Conchyl. xv, p. 416.—*Lepades vertice integro, margine lacero, ovatœ, costoso-plicata,* etc. MARTINI, Conchyl. Cab. i, p. 91, t. 8, f. 66 ; also *Lepas magna, vertice integro acuto, albo,* etc., etc., *t. c.,* p. 117.—*P. rouxi* PAYR., Moll. de Corse, p. 90, t. 4, f. 1, 2.—*P. pyramidata* LAM., An. s. Vert. vi, p. 327.—DELESSERT, Rec., t. 22, f. 3.—*P. ferruginea var. pyramidata* WEINKAUFF, Conchyl. des Mittelm. p. 401.

The strong rounded ribs, deeply crenulating the margin, distinguish this from other European species.   The interior is faint bluish and slightly opalescent outside of the muscle-scar ; inside of it there is a distinctly defined, opaque white callus.   The ribs number from 44 to 50 in all.

6

There is no warrant whatever for the use of the name "*contoso-plicata* Martini" for this shell. There is not the slightest pretension to or attempt at binomialism or the use of generic names in the first volume of Martini.

De Gregorio has described the following varieties : *sitta, imperatoria, percostata, ficarazzensis* (Bull. Soc. Mal. Ital. x. p. 120, 124.)

P. VULGATA Linné. Pl. 10, figs. 1–6.

Shell solid, oval, conical, the apex a little in front of the middle ; slopes nearly straight ; surface sculptured by numerous radiating ribs (often subobsolete), the spaces between the ribs having radiating striæ. Color varying from whitish to pink, yellow, slate, olive, or black, the ribs generally lighter.

Interior somewhat opalescent in dark specimens, usually yellowish and showing faint rays around the edge, the central area varying from white to dark-brown. Length 44, breadth 37, alt. 17 mill.

Lofoten Is., Norway, to Spain.

P. vulgata L., Syst. Nat. xii, p. 1258.—FORBES & HANLEY, Hist. Brit. Moll. ii, p. 421, t. 61, f. 5, 6.—JEFFREYS, Brit. Conch. iii, p. 236 ; v, t. 57, f. 1–4 (with varr. *elevata, picta, intermedia, depressa, cærulea*).—HIDALGO, Mol. Mar. Esp. t. 52, f. 1–8 ; t. 53, f. 7, 8.— DALL, Amer. Journ. Conch. vi, p. 268, t. 15, f. 23 (anatomy).— SARS, Moll. Reg. Arct. Norv., p. 118, t. ii, f. 7a. 7b. (dentition).— HARVEY, Trans. Roy. Soc. Edinb. xxxii, pt. 3, p. 601–636, 1885 (anatomy and histology).—? *P. radiata* PERRY, Conch., t. 43, f. 1.

The common Patella of northern Europe is the typical *vulgata* of Linné. It is more elevated than the Mediterranean shells, but some specimens of the latter can scarcely be separated specifically. The species is excessively variable : The forms recognized by Jeffreys occurring on the English coast are as follows :

Form *elevata* Jeffr. Much smaller, rounder and higher.

Form *picta*. Smaller and thinner ; with alternate rays of reddish and dark blue.

Form *intermedia* Knapp. Smaller, flatter and oval, with finer ribs and an orange crown ; inside golden-yellow or tinged with flesh-color (occasionally cream-color) in the center, and beautifully rayed toward the margins (Ann. Mag. N. H. xix, 1857, p. 211).

Form *depressa* Pennant (pl. 11, figs. 24, 25, 26). Much depressed, more oblong than the usual form ; ribs finer but sharp ; apex more anterior ; inside porcellanous with a pale orange head scar or spatula. *P. athletica* F. & H. is a synonym. This form is thicker and more coarsely sculptured than the var. *aspersa* Lam.

P. CÆRULEA Linné. Pl. 10, figs. 7–12.

Shell depressed, thin, spreading, usually more or less distinctly 6 or 7 angled; riblets rather fine and notably unequal.

Mediterranean and Adriatic Seas; Madeira; Azores; Canaries.

P. cærulea L., *l. c.*, p. 1259.—HANLEY, Sh. of L., p. 421.—PHIL., Enum. Moll. Sicil. i, p. 109, t. 7, f. 5.—REEVE, Conch. Icon., f. 28. —HIDALGO, Mol. Mar. Esp., t. 50, f. 5, 6; t. 51, f. 1, 2.—BUQ. DAUTZ. & DOLLF., Moll. Mar. Rouss., p. 471, t. 58, f. 1–7.—*P. fragilis* PHIL., Enum. i, p. 40, t. 7, f. 6.—*P. subplana* P. & M., Galerie de Douai, i, p. 524, t. 37, f. 3, 4.—*P. cærulea v. subplana* BUQ. DAUTZ. & DOLLF., Moll. Rouss., p. 473.—*P. tarentina* LAM., *not* v. SALIS.—*P. scutellaris* BLAINVILLE, *not* LAM.

P. aspera LAM., An. s. Vert. vi, p. 327.—*P. bonnardi* RVE., *not* PAYR.

P. tarentina v. SALIS, Reise ins Koenig. Neapel, p. 359, t. 6, f. 2.— *P. bonnardi* PAYR., Moll. de Corse, p. 89, t. 3, f. 9–11.

P. crenata GMEL., Syst., p. 3706.—ORB., Moll. Canaries, p. 97, t. 7, f. 6–8.—DROUET, Moll. Mar. Açores, p. 40.

Separated from *P. vulgata* mainly on account of its more expanded, depressed, generally thinner shell and more southern range. I am wholly inclined to believe that the line of separation is artificial, and that the two species *vulgata* and *cærulea* fade into one another.

The variations of the genuine *cærulea* are numerous including the following forms:

Form *fragilis* Phil. Shell thin, the radiating striæ very fine.

Form *intermedia* B. D. & D. Intermediate between the regularly oval and the polygonal forms.

Form *adspersa* B. D. & D. Dotted with white on a greenish-gray ground.

Form *subplana* Pot. & Mich. (figs. 7, 8). Large, thin, pentagonal, the apex quite anterior. This is *P. scutellaris* of Blainville, Reeve, and others. As mutations under it rank form *stellata* B. D. & D., having the angles prolonged, star-like; form *cognata* B. D. & D., having the pentagonal form of *subplana* and the rugose sculpture of *aspera*.

Var. ASPERA Lamarck.   Pl. 11, figs. 20, 20a, 21, 22, 23; pl. 53, figs. 4, 6.

Solid, depressed, the growth-lines rising into more or less prominent scales on the conspicuous ribs.

Mediterranean and Adriatic Seas.

Form *tarentina* von Salis.   Pl. 53, fig. 6.   Conspicuously rayed with brown; nearly smooth.

Form *spinulosa* B. D. & D. (pl. 53, fig. 4).   Ribs spinose.

Additional names applied to forms belonging to the *vulgata* and *cærulea* stock are forms *comina, depressaspera, macrina, albula* and *cimbulata* De Greg., Bull. Soc. Mal. Ital. x, 1884, pp. 122, 123; *P. taslei, ordinaria, goudoti, servaini* and *teneriffæ* J. Mabille, Bull. Soc. Philomathique de Paris, 1887–1888, pp. 78–81.

Var. CRENATA (Gmelin) Orbigny.   Pl. 54, figs. 12, 13, 14.

Depressed, irregularly oval, having numerous rather low riblets, over which small granules are scattered more or less closely. Yellowish-brown or tawny outside; the inside usually bluish, more or less iridescent, white in the middle.

Azores and Canary Is.

Var. LOWEI d'Orbigny.   Pl. 53, figs. 7, 8, 9, 10, 11 ; pl. 29, figs. 44, 45, 46.

Depressed, spreading, angularly ovate, solid, the surface very densely ribbed, the ribs unequal, scaly, especially toward the margins.   Color varying from dull brown or rust red to blackish-brown; sometimes rayed with white.   Edge of the shell denticulate, the projections compound, foliated.

Inside varying from white to deep blue, iridescent, having a fibrous texture; the central area generally white, and much thickened, callous.

Madeira, Azores, Cape Verde and Canary Is.

P. lowei ORB., Hist. Nat. Canaries, Moll., p. 97, t. 7, f. 9, 10.— DAUTZENBERG, Mém. Soc. Zool. France, iii, p. 161, 1890.—DROUET, Moll. Mar. Açores, p. 40.—*P. azorica* NUTT., *teste* JAY, Catal., 4th edit., 1852, p. 100, no. 2798.

More spreading and angular than var. *aspera*, and the marginal denticulations are foliated.   Figure 43 of pl. 29 is not characteristic.

The other figures are drawn from specimens before me. Dr. *H. Simroth* (Zur Kenntniss der Azorenfauna, in Archiv für Naturgeschichte 54th year, 1888, p. 215, Apr., 1889) ranks the Azores *Patella* under the single species *aspera*, with several varieties as follows :

"*Patella aspera* Lam.

a. *typical form,*=*P. lowei* Orb., *P. baudoni* Drouet, *P. spectabilis* Drouet, not Dkr.

b. *P. moreleti* Drt., *P. crenata Gm.* of Orb., in Moll. Canar., and probably *P. gomezi* Drt.

c. *var. simrothi* v. Martens. Narrower, oval, flat. Ribs weaker, more rounded than carinated, either entirely smooth or having distinct scales. Margin but little crenated. Color of the outside pretty regular dark reddish-brown ; inside also pretty dark, obscure violet, sometimes more reddish, sometimes more dark blue, the central area bluish-white or gray-blue, sometimes with an admixture of yellow ; rarely having distinct dark rays inside. The margin is often somewhat horizontally dilated. Apex at the front ⅔. Length 41, breadth 30, alt. 14 mill. ; length 32, breadth 24, alt. 10 mill. Azores. Approaches *P. cœrulea* L. but evidently deserves a separate name.

d. *var. accedens* ad *lusitanicum* Gmel. Higher, broader more bluntly elevated, ribs and border similar to the foregoing. Colored outside and inside with broad dark, defined rays on a light ground. Central area lead-gray, rarely yellowish or reddish. Only small specimens, length 25 breadth 21, alt. 11 mill., apex at ⅖ of the length. The *P. nigrosquamata Dkr.* of Drouet is perhaps this, but without rays."

P. MORELETI Drouet. Pl. 56, figs. 27, 28.

Shell subdepressed, very rugose, ribbed, the ribs scaly, scarcely solid ; brownish-green outside : inside brownish or reddish, iridescent, with a white spot at the summit. Apex acute. Aperture ovate, crenulated. Length 40, breadth 30, alt. 12 mill. (*Drouet.*)

Fayal, Azores.

P. moreleti DR., Moll. Mar. Açores, p. 42, t. 2, f. 10, 11, 1858.

Considered a variety of *P. aspera* by Simroth.

P. GOMESII Drouet. Pl. 54, figs. 17, 18.

Shell large, subdepressed, rugose, ribbed-plicate, rather solid; outside grayish-brown or rufescent; inside shining, brown, pearly; apex situated at the front third of the length, obtuse; aperture oval, entire. Length 50–60, breadth 50–53, alt. 12–15 mill. (*Drouet.*)

Bay of San Lourenzo, Santa Maria, and Pico, Azores Is.

P. gomesii DROUET, Mollusques Marins des Iles Açores, p. 39, t. 1, f. 6, 7, 1858.

More depressed and less elongated than *P. candei*, the ribs more conspicuous, the summit more obtuse.

Referred by Simroth to *P. aspera var. moreleti* Drt.

P. BAUDONII Drouet. Pl. 54, figs. 15, 16.

Shell large, subelevated, coarsely ribbed, plicate, solid, thick; outside grayish-green, inside white; vertex subacute, submedian; aperture oval, a little crenated.

Length 60, breadth 50, alt. 25 mill. (*Drouet*)

Santa Maria and Pico, Azores.

P. baudoni DR., Moll. Mar. Açores, p. 41, t. 2, f. 8, 9, 1858.

This seems to be very closely allied to *P. ferruginea* Gmel. It is referred by Simroth to *P. aspera.*

P. CANDEI d'Orbigny. Pl. 55, fig. 22, 23, 24.

Shell elevated, conical, thick, smooth or irregularly roughened; ovate, margin entire. Inside buff, bluish in the middle; outside pale yellow. Length 67, breadth 58, alt. 27 mill. (*Orb.*)

Canaries.

P. candei ORB., Hist. Nat. Canaries, ii, 2d part Mollusques, p. 98, t. 7, f. 11, 12, 1844.—REEVE, Conch. Icon., f. 34, 1854.

P. CITRULLUS Gould. Pl. 28, figs. 39, 40, 41.

Shell sub-diaphanous, thin, sub-conical, moderately elevated, summit prominent; apex anterior, acute, feebly incurved, usually somewhat eroded; a great number of faintly elevated lines, studded with fine tubercles or asperities, radiate from it, and become obsolete about half way towards the margin. Striæ of increment coarse and irregular, overlaying each other, so as to give the shell a rude, concentrically squamose aspect externally; disk nearly oval, a little narrowed anteriorly; margin very thin and sharp, finely and irregularly undulated. External color a dusky olive-green, with a shade of brown showing through it, ornamented with concentric, undulating lines of obscure white. Interior greenish-white, with bright iridescent reflections; a slight spatulaform deposit at the fundus, bluish at the edges and forepart, passing into greenish towards the middle and posterior portions. (*Gld.*)

Length 1⅜, breadth 1⅛ inch.

Funchal, Madeira Is.

P. citrullus GLD. Proc. Bost. Soc. N. H., ii, p. 149, July, 1886; U. S. Expl. Exped. Moll. & Sh. 335, f. 448.

This shell resembles somewhat the skin of a cucumber externally. The radiating striæ occupy the upper half of the shell, and the lower half is somewhat imbricated by the stage of growth. It is somewhat like *P. Candei* D'Orb. (*Gld.*)

I have not seen this species. It should be compared with *P. cærulea* var. *crenata*, and with *P. candei*.

P. LUSITANICA Gmelin. Pl. 11, figs. 15, 16, 17, 18, 19.

Shell solid, conical, rounded-oval, the apex elevated and slightly in front of the middle; front slope straight, posterior slope slightly convex. Surface dull, lusterless; having close unequal, granose radiating riblets. Ashen-white, with blackish rays wider than the white ones, the granules on the ribs black.

Interior rayed with brown or purplish-black on a lighter ground, the central area white, brown or blue-black, often surrounded by a yellow stain. Length 35, breadth 29, alt. 16 mill.

Mediterranean and Adriatic Seas; Atlantic coast of Portugal and S. W. France (Gironde); Madeira.

P. rustica LINN., Syst. Nat. x, p. 783, *teste* HANLEY, Shells of L., p. 427 (Not *P. rustica* LINN., Mus. Ulricæ, nor of REEVE, Conch. Icon., nor of MENKE, Moll. Nov. Holl. Spec.).—*P. lusitanica* GMEL., Syst. p. 3715.—PHIL., Enum. Moll. Sicil. i, p. 110.—HIDALGO, Mol. Mar. Esp. 51, f. 3, 8.—BUQ. DAUTZ. & DOLLF., Moll. Mar. Rouss. i, p. 469, t. 57.—*P. granularis* v. SALIS *et al.*, not Linn.—*P. nigropunctata* REEVE, Conch. Icon. f. 57.—*P. piperata* GLD., U. S. Exped. Moll. p. 338, atlas f. 449.

The conical form and dark or black granules upon the radiating riblets readily distinguish this species.

P. piperata Gould (pl. 29, figs. 50–52) is identical. The *P. rustica* of Linné (Syst. Nat. x) is, according to Hanley, who has studied Linné's type, the same as *lusitanica*; but Linné afterward described another and entirely different shell under the same name, and as his original description is insufficient, the name *rustica* had better be dropped entirely.

The conical, compact form and subgranose riblets, dotted with blackish, are characteristic. The species may perhaps be found to intergrade with *P. guttata* Orb., but proof of this is lacking at present.

P. GUTTATA d'Orbigny.   Pl. 56, figs. 29, 30, 31.

Shell elevated (the young depressed), conical, thick, unequally ribbed, the ribs alternately large and small, longitudinally tuberculate, the tubercles black; interior grayish, yellowish-red in the middle.   (*Orb.*)

Length (adult specimen) 54, breadth 46, alt. 35 mill.

Length (young specimen) 23, breadth 19, alt. 5 mill.

> *Teneriffe and Grand Canary Is., Canaries.*

P. guttata ORB., Hist. Nat. Canaries, Moll., p. 98, t. 7b, f. 13–15, 1844.—REEVE, Conch. Icon., f. 91.—DAUTZENBERG, Mém. Soc. Zool. Fr. iii, p. 161, 1890.—*P. frauenfeldi* DUNKER, Verh. k. k. zool.-bot. Ges. in Wien, xvi, p. 914, 1866.—FRAUENFELD, Reise der Oesterreichischen Fregatte Novara, Zool. Theil, ii, Moll., p. 15, t. 2, f. 26.

The black nodules upon the ribs and the red stained interior are prominent and characteristic marks.   It is somewhat allied to *P. granularis, natalensis*, etc.

P. nigrosquamosa Dkr. is probably a variety or synonym of this species.   Von Martens (Zool. Rec. iii, p. 188) having examined part of the original specimens of *P. frauenfeldi*, considers them identical with *guttata*.   The locality " Madras " is an error.

P. NIGROSQUAMOSA Dunker.   Pl. 13, figs. 57, 58, 59.

Shell ovate, convex-conic, whitish with large and small scaly radiating ribs; scales erect, blackish. Apex subcentral, rather acute. Margin crenulated. (*Dkr.*) Length 19 mill.

> *Horta, Fayal Is., Azores.*

P. nigrosquamosa Dkr., Zeitschr. f. Mal. 1846, p. 25.—*P. nigro-squamosa var. minor* Dkr., Ind. Moll. Guin. Infer. p. 41, pl. 7, f. 4–6.

We must retain this name for the Azores form for which it was originally proposed. Dunker subsequently included the larger Cape species, *P. natalensis.*

Compare *P. guttata* Orb. of which this may prove to be a variety.

P. RANGIANA (Valenciennes) Rochebrune. Pl. 58, figs. 42, 43.

Shell ovate, depressed-convex, rufous; vertex submucronate, usually eroded, situated at ⅔ of the length; having larger and smaller radiating broad, very scaly ribs, scales subimbricating, obtuse, lenticular; margin undulating; interior bluish, silvery-pearly, rayed with bands and spots of purplish, the center spatulate, pale orange. Length 44, breadth 36, alt. 19 mill. (*Rochebr.*)

Porto Praya, Cape Verdes.

P. rangiana (Valenc. *ms.*) ROCHEBRUNE, Bull. Soc. Philomathique de Paris, 7th Ser., vi, p. 29, 1882; Nouv. Arch. du Mus. 2d Ser., iv, p. 267, t. 18, f. 7, 1881.

P. GUINEENSIS Dunker. Pl. 12, figs. 34, 35, 36, 37, 38.

Shell oval, sometimes elliptical, subelevated, rather solid, concentrically striate and subrugose; furnished with close, unequal subnodose radiating ribs, in part obsolete. Buff-orange, rayed and variegated with brown. Apex projecting, inclined forward, submamillar, situated at the front ¼ of the length.

Interior buff-orange, center milky and lurid. Margin acute, slightly crenulated, nearly simple, Length 27 mill. Ratio of length, breadth and alt. $= 100:88:30$. (*Dkr.*)

Loanda, Guinea.

P. guineensis Dkr., Ind. Moll. Guin. Infer. p. 40, t. 7, f. 19, 20, 21, 1, 2, 3.

P. SPECTABILIS Dunker. Pl. 12, figs. 45, 46, 47.

Shell ovate, subdepressed, solid, buff or dull white, rayed with reddish; apex somewhat obtuse, situated at ⅔ the length, or the space in front bearing to the space behind the apex the ratio of 3 to 4. Having numerous radiating ribs, which are rugulose, subnodose, subimbricated toward the margin. Interior bluish, the center white; margin irregularly folded, crenulated, scarcely angular. Length 43, breadth 34½, alt. 12 mill. A large example measures 62 mill. in length. (*Dkr.*)

Loanda, Guinea.

P. spectabilis DKR., Ind. Moll. Guin. Infer., p. 39, t. 6, f. 7–9.

Should be compared with *P. cœrulea* var. *aspera.*

P. SAFIANA Lamarck.   Pl. 55, figs. 19, 20, 21.

Shell ovate-oblong, convex, variable; having equal radiating flattened white ribs, the interstices brown; apex subacute, inflexed. It is of grayish whitish above, and rayed between the ribs with yellowish or slightly brown rays. The internal border is of a bluish nacre. (*Lam.*)

Ocean coast of Morocco.

P. safiana LAM., An. s. Vert. vi, p. 329, 1819.—DELESSERT, Rec. de Coq. t. 22, f. 2.

The following species, *conspicua* Ph., is probably identical with this forgotten Lamarckian shell.

P. CONSPICUA Philippi.   Pl. 56, figs. 25, 26, 26.

Shell rather thick, ovate, a little broader behind than before, with about 34 flat, somewhat sharply angular, coarse ribs, and about double that number of weaker riblets in their interspaces; the border is irregularly toothed and crenated by these ribs. The apex lies at the front third of the length; the front slope is straight, back slope convex. The color is whitish, with blackish-brown interrupted rays, here and there connected by transverse striæ. Inside, the outer portion is whitish, showing the external black rays through the shell; in the center more or less reddish-yellow. Sometimes the reddish-yellow color includes the central area and the muscle-impression, sometimes the muscle-impression is the darkest reddish-yellow, and the centrum itself lighter. (*Phil.*)

Length 80, breadth 62, alt. 24 mill.

Gaboon. Guinea.

P. conspicua PHIL., Abbild. iii, p. 71, t. 3, f. 1, October, 1849.— DKR., Ind. Moll. Guin. Infer., p. 43.—? REEVE, Conch. Icon., f. 12.

I have not identified this species with certainty. The original figures and description are given.

It is doubtful whether Reeve's figures represent this species. I have copied them on pl. 21, figs. 47, 48.

P. LUGUBRIS Gmelin.   Pl. 57, figs. 32, 33, 34, 35; pl. 12, figs. 39, 40, 41, 42, 43, 44.

Shell conical, short-oval, solid; the apex situated in front of the middle; slopes slightly convex; sculptured with numerous (34 to

37) strong radiating ribs, most of which are compound, as if formed by the coalescence of several smaller ribs. Color dull black, sometimes rayed with white.

Interior blue, the central area either blue or white; border crenulated. Length 60, breadth 50 alt. 20 mill.

Loanda and Benguela, Guinea ; Cape Verde Is.

P. lugubris GMELIN, Syst. p. 3705.—REEVE, Conch. Icon. f. 32, 1854.—DUNKER, Ind. Moll. Guin. Infer. p. 38, t. 7, f. 9–11, 22–24.

The uniform black color is sometimes relieved by light rays, visible inside, and sometimes the whole central area is white, the black rays not extending to the apex. These color-marks are best seen by looking *through* the shell at a strong light. The ribs are usually seen to be *compound*, or partially split into several smaller riblets; but of the principal ribs there are generally about 34.

P. PLUMBEA Lamarck. Pl. 24, figs. 11, 14, 15; pl. 57, figs. 38, 39.

Shell low-conic, oval or elliptical, rather solid; apex at the anterior two-fifths of the shell's length; slopes convex. Surface sculptured with numerous riblets, sometimes narrow, irregular and close, sometimes separated. Color dull black, sometimes rayed with grayish, the rays scarcely perceptible outside.

Interior blue, the central area elongated, white, often clouded or stained with rich brown. Margin crenulated.

Length 53, breadth 42, alt. 15 mill.

Senegal ; St. Helena.

P. plumbea LAM., An. s. Vert. vi, p. 328, 1819.—DESH., *l. c.* vii, p. 530.—REEVE, Conch. Icon. f. 5 and 46.—E. A. SMITH, P. Z. S. 1890, p. 296.—*P. cærulea* QUOY & GAIM., Voy. de l'Astrol. Moll., iii, p. 342, t. 70, f. 4–6.—*P. cyanea* LESSON, Voy. de la Coquille, ii, p. 417.—*P. canescens* RVE., Conch. Icon. f. 103.

This species is extremely variable in sculpture. The ribs are narrow, rather separated, but often increased in number and closeness by the secondary ribbing which transforms each rib into *three*, the middle one larger. The ribbing is finer than in *P. lugubris*, the shell is more elliptical, more depressed, and the central spatula of the interior is longer and narrower.

The variations exhibited by the series before me cause me to regard *P. safiana* Lam. and *conspicua* Phil. as close allies, possibly varieties of this species.

The form called *P. canescens* by Reeve is figured on pl. 57, figs. 36, 37. It has quite a different appearance, but I am disposed to believe that Mr. Smith is justified in placing it in the synonymy.

Var. VATHELETI Pilsbry. Pl. 57, figs. 40, 41, 42, 43.

Shell low-conic, ovate-rectangular, solid; sculptured with numerous unequal riblets, and having about nine larger but often indistinct ribs or angles, distinguished by white stripes, giving a more or less polygonal outline to the shell. The radiating ribs and riblets are closely cut or crenulated by concentric striæ. The principal ribs are white, the intervals black. Upper part of the cone eroded, grayish-white.

Interior whitish, somewhat stained with yellow, blotched around the margin with black, the central area either coal-black or marbled black and white.

Length 28, breadth 22, alt. 7 mill.

Senegal.

These small shells are evidently allied to *P. plumbea*, but differ in contour, in the very short, broader central area of the interior, etc. They were collected by the Abbé A. Vathelet.

P. ADANSONII Dunker. Pl. 12, figs. 30, 31, 32, 33.

Shell oblong-ovate, subelliptical, moderately elevated, sculptured with about 80–100 unequal ribs; whitish, marbled and striated with black, brown and olive. Apex situated in front of the anterior third of the length. Interior bluish, the central area white and rufescent, sometimes rather flesh-colored or liver-colored; toward the edge rayed with reddish. Margin subcrenate-dentate. (*Dkr.*)

Length of large specimen 50, breadth 41, alt. 17 mill.

Loanda, West Africa.

P. adansonii DKR., Ind. Moll. Guin. Infer., p. 42, t. 6, f. 10–15, 1853.

This species is evidently closely allied to *P. plumbea* Lam., differing mainly in the varied coloring of the exterior. The number of riblets is very variable; one specimen has 120. There is a small riblet on each side of the larger ones, as in *conspicua*, *plumbea*, and many other species from this region.

P. KRAUSSII Dunker. Pl. 13, figs. 54, 55, 56.

Shell ovate, thin, rather depressed, pale brown, subcorneous, radiately ribbed and transversely striated, ribs unequal, subundulating; apex acute, curved over, somewhat hooked, situated nearly at ⅔ of the length. Margin acute, obsoletely plicate and crenulated beneath. Interior somewhat hoary and bluish, the center dull white and yellowish. Length 31, breadth 22, alt. 8 mill. (*Dkr.*)

Loanda, West Africa.

P. kraussii DKR., Ind. Moll. Guin. Infer., p. 42, t. 6, f. 4–6, 1853.

There are 100–120 radiating ribs; the intervals between the larger ribs are occupied by 2 or 3 smaller riblets. The larger ribs are scaly and nodulous; and the whole surface is very delicately lineated.

P. COMPRESSA Linné. Pl. 61, figs. 68, 69, 70.

Shell thin, *narrow, oblong, the sides compressed* and parallel ; conical, the apex in front of the middle and curving forward. Covered with close unequal radiating riblets ; *dull straw-colored*, the young often finely dotted, spotted or rayed with bright crimson, pink or orange often marked with opaque-white dots or triangles.

Interior white, or in the young marked like the outside. Edge of the shell even, *the ends elevated.*

Length 94, breadth 45, alt. 35 mill.

Cape Good Hope.

P. compressa LINN., Syst. x, p. 783.—KRAUSS, Südaf. Moll. p. 50.—Q. & G. Voy. de l'Astrol., iii. p. 338, t. 70, f. 1.—REEVE, Conch. Icon. f. 13a, and of authors.—*P. miniata* BORN, Test. Mus. Cæs Vindob. p. 420.—LAM., An. s. Vert. vi, p. 333.—KRAUSS, Die Südaf. Moll. p. 51.—*P. umbella* GMEL., Syst. p. 3706.—LAMARCK, An. s. Vert. vi, p. 327.—REEVE, Conch. Icon. f. 17.—*P. sanguinolenta* GMEL., Syst., xiii, p. 3716, no. 130.—*P. sanguinalis* RVE., Conch. Icon. f. 95.

This species is readily known by its compressed sides, elevated end-margins, and straw-yellow color. It is occasionally found variegated with red on the upper part of the cone, and the young are almost always so marked.

Dead specimens have been found at St. Helena, doubtless drifted thither on seaweed. (See Smith, P. Z. S., 1890, p. 248.)

The typical form of this species is never found living on shore. It lives upon large seaweeds, as one might readily tell by the form of the shell. When living on rocks it develops into the form known as *P. miniata*. Specimens which have changed their stations and show a corresponding abrupt change of sculpture are not very infrequent. Parallel mutations occur in *Acmæa pelta*, *Patella granularis*, *Patina pellucida*, etc., etc.

Var. MINIATA Born. Pl. 26, figs. 22–27.

Shell thin, varying from depressed and broadly ovate to conical and narrowly ovate; sculptured with numerous acute unequal radiating riblets, more or less destinctly decussated by growth-striæ. Riblets white or yellowish, interstices occupied by red rays; the young speckled and blotched with red (rarely brown or purple-black).

Interior having an opaque-white central area, the outer portion transparent and showing the color-rays.

Length 70, breadth 58, alt. 13 mill. (normal.)

Length 58, breadth 42, alt. 18 mill. (more elevated specimen; figs. 23, 24.)

There is great variation in the form, still greater in the sculpture of this shell. Some young specimens are nearly smooth, having fine, subequal radiating riblets, scarcely decussated, whilst others have the riblets distinctly cut into close, compressed beads by the concentric sculpture.

In young shells the apex is much nearer the anterior end than in adults.

P. ELECTRINA Reeve. Pl. 18, figs. 33, 34.

Shell orbicular, attenuated in front, rather depressed; sharp at the apex; radiately densely ridged, ridges rough, irregular, bluntly squamate. Light fulvous, rusty about the apex and between the ridges. Interior transparent white. (*Rve.*)

Australia.

P. electrina RVE., Conch. Icon., f. 55. Dec., 1854.

Of a transparent texture, stained with amber rust about the apex and between the ridges, the color showing conspicuously through in the interior. (*Rve.*)

Compare *P. miniata* and *P. lowei.*

Section SCUTELLASTRA H. & A. Adams, 1858.

Scutellastra ADS., Genera Rec. Moll. i, p. 466, types *P. gorgonica* Humph., *pentagona* Born, *plicata* Born.—*Olana* ADS. *l. c.,* type *P. cochlear* Gmel.

Inner layer of the shell opaque, porcellanous, not iridescent nor fibrous in texture.

Animal similar to *Patella*, except that a small rhachidian tooth is frequently developed.

Distribution, S. Africa to Central Pacific.

I have separated this group from *Patella s. s.* mainly on account of the different texture of the shell. Although this distinction has not heretofore been noticed, I am confident that it is constant and of sufficient systematic value to warrant the course here taken.

(1) GROUP OF P. BARBARA.

Erect, oval shells, with numerous ribs or riblets.

P. ARGENVILLEI Krauss. Pl. 22, figs. 15, 16; pl. 58, fig. 44.

Shell large, solid, elevated-conical, ovate, apex in front of the middle, slopes nearly straight. Surface closely sculptured with numerous (80–100) crowded, obtuse radiating riblets, obscurely alternating in size, and roughened by the low, scaly growth-lines. Color blackish, having concentric lighter zones.

Interior white, stained at the muscle-scar with brown or yellowish-brown, the edges of the central area well-defined, laciniate, stained: border brown, closely crenulated by small teeth arranged in pairs.

Length 81, breadth 62, alt. 43 mill.

Table Bay, South Africa.

P. argenvillei KRAUSS, Die Südafric Moll., p. 49.—REEVE, Conch Icon., f. 20.—? *Lepas écaillé* ARGENVILLE, La Conchyl., p. 504, t 3, f. G.

P. NEGLECTA Gray. Pl. 20, figs. 41, 42; pl. 58, figs. 40, 41.

Shell large, solid, elevated-conical, elliptical or ovate; apex at about the front third; slopes nearly straight. Surface sculptured with coarse, irregularly subnodose, unequal radiating riblets. Ribs whitish, interstices blackish brown.

Interior white, tinged with flesh-color, having some yellowish-brown clouds or stains in the central area; muscle-scar distinct, light-buff or flesh-tinted. Edge of the shell crenated, conspicuously marked with black-brown blotches, mostly in pairs.

Length 95, breadth, 68, alt. 40 mill.

Length 106, breadth 80, alt. 41 mill.

King George's Sound, Mistaken Island, and Swan River, Australia.

? *P. rustica* LINN., Mus. Lud. Ulricæ p. 694, (not of Linn. Syst. Nat. x, = *P. lusitanica* Gm., *q. v.*)—*P. rustica* L., MENKE, Moll.

Nov. Holl. p. 33, 1843 ; and also Zeitschr. f. Malac. 1844, p. 62.—
P. pileus MKE., *mss.*—*? P. indica* GMEL., Syst. Nat. xiii, p. 3716,
founded on Gualtieri, Testarum, t. 8, f. E, and Martini, Conch.
Cab. i, p. 106, t. 7, f. 49.—*P. melanogramma?* SOWERBY, Genera,
Patella, f. 1 (good!) ; not *P. melanogramma* Gmel., Syst. xiii, p.
3706, no. 74.—*P. neglecta* GRAY, in Capt. King's Survey of the
Inter-tropical and Western Coasts of Australia, ii, appendix, p.
492, 1827.—*P. zebra* REEVE, Conch. Icon. f. 7, Oct. 1854.

This large species is much more strongly ribbed than *P. argent-
villei*. The ribs are very unequal. The position of the apex is
more anterior in my specimens than in Reeve's figures, in one specimen being decidedly in front of the anterior third, nearly reaching
the fourth. The ribs are rudely nodular in young or half grown
shells, but become obsolete and eroded with age.

Notes on synonymy.—That this species is not the *P. rustica* of
Linnæus' Systema x, is obvious (see under *P. lusitanica*, this volume).
Whether it is the *rustica* of his later publications (Mus. Lud.
Ulricæ, p. 694, etc.) or of Gmelin, is a useless question into which
we need not enter, but with Menke, I am disposed to believe that it
is. Reeve's *P. rustica* has nothing to do with this species, being
either a large *P. lowei* or an immature *P. patriarcha*. The *P. indica*
of Gmelin, founded upon Gualtieri's figure and Martini's embellished
copy of it, is very doubtful at best. Sowerby gave a most excellent
figure under the name *melanogramma*, but it is not the shell so
named by Gmelin. Gray fixes the identity of his *P. neglecta* by
stating that it is the *P. melanogramma* of Sowerby's Genera, not of
Gmel.

P. BARBARA Linné.   Pl. 59, figs. 50, 51, 52, 53, 54, 55 ; pl. 15, figs.
1, 2.

Shell rather large, depressed or conical, ovate ; apex central ;
slopes nearly straight. Sculptured with numerous elevated, acute
narrow riblets, which bear conspicuous narrow vaulted spines.
White or tinged with brown, the spines usually tipped with brown.

Interior white, either immaculate or having the central area
stained with light orange-brown. Margin strongly toothed, having
a colorless border. Length 72, breadth 60, alt. 27 mill.

Habitat unknown.

P. barbara LINN., Syst. Nat. x, p. 782.—HANLEY, Shells of
Linnæus, p. 418.—LAM., An. s. Vert. vi, p. 325.—*P. plicata* BORN,

Mus. Cæs. Vindob., t. 18, f. 1.—Reeve, Conch. Icon., f. 16.—*P. barbata* Lam., An. s. Vert. vi, p. 326.—Delessert, Rec. de Coq., t. 21, f. 1.—*P. spinifera* Lam., An. s. Vert. vi, p. 326.—Delessert, Rec., t. 21, f. 2.—?? *P. cypria* Gmel., Syst. xiii, p. 3698.

In all the variety of forms and names in which this species masquerades, it may be known by the acute, high, compressed ridges, which bear vaulted or sometimes solid spines, usually touched with brown at their tips. The ribs are unusually variable in number, 24 to 30 being developed on moderate sized individuals, not counting a few small interstitial riblets. The spines are rarely as numerous as Reeve's figures show. These are reproduced on pl. 15, figs. 1, 2.

An elevated, conical form is figured on pl. 59, fig. 55.

A form which may be known as *var.* ovalis is figured on pl. 60, figs. 56, 57, 58. It is ovate and has about 41 ribs. Interior pure white.

Length 95, breadth 70, alt. 31 mill.

(2) Group of P. stellæformis.

Shell having coarse, unequal ribs or riblets.

P. pica Reeve. Pl. 22, figs. 9, 10, 13, 14; pl. 59, figs. 47, 48, 49; pl. 26, figs. 28, 29.

Shell solid, depressed, apex a little anterior, ovate, broad behind, more or less narrowed in front; having numerous (about 21) rude angular radiating ribs, and more or less obviously radiately striate. Whitish or ashen, irregularly blotched with black.

Interior white, the central area sometimes stained with yellowish or brown. Edge crenated, having a narrow border which is usually whitish-buff dotted and blotched with black, but sometimes lacks all dark markings.

Length 43, breadth 31, alt. 10 mill.
Length 47, breadth 37, alt. 10 mill.

Mauritius and Reunion.

P. pica Reeve, Conch. Icon., f. 45, 1854; also f. 68.—*P. chitonoides* Reeve, f. 52.—Desh., Moll. Réunion, p. 43.—*P. moreli* Dh., Moll. Réunion, p. 43, t. 6, f. 13.—*P. levata* Dh., l. c., p. 44, t. 6, f. 14.—? *P. dentata* Dufo, Ann. Sci. Nat. 1840, p. 204.

The outline is somewhat spoon shaped, approaching that of *P. cochlear*. The ribs are usually subequal, and about 21 in number, but sometimes they are very irregular, as in the figures on pl. 59.

7

In the form called *chitonoides* Rv. (pl. 26, figs. 28, 29) the ribs are somewhat more numerous, and the entire surface is purple-black.

I am unable to find differential characters in the *P. moreli* of Deshayes. I have copied the original figure on pl. 58, fig. 45, representing an immature specimen. The same is true of *P. levata* Dh. (pl. 59, fig. 46).

This species seems to be more than usually encrusted with calcareous growths, algæ, etc. The specimens before me are from Mauritius, collected by Robillard.

P. EXUSTA Reeve. Pl. 24, figs. 9, 10.

Shell ovate, a little attenuated in front, flatly convex, rather spread; apex inclined anteriorly; radiately ribbed, ribs with the surface rude, irregularly, obscurely prickly-scaled, interstices obsoletely latticed, with rather distant concentric ridges; burnt-black, red-tinged, interior marble-white; edge remotely denticulated, purple-black.

An extremely characteristic species, marble-white within, reddish-burnt black without, but of singular rude irregular, obsoletely latticed sculpture. (*Rve.*)

Habitat unknown.

P. exusta RvE., Conch. Icon., f. 35, Oct., 1854.

This may prove to be the same as *P. pica* Reeve. If so the name *exusta* will take precedence.

P. FUNEBRIS Reeve. Pl. 60, figs. 59, 60, 61.

Shell ovate, slightly attenuated in front, elevately convex; smooth, rayed with tubercled ribs, tubercles swollen, sometimes rather distant. Dull black, rusty-white at the apex. Interior opaque white, sometimes rust-tinged. (*Rve.*)

Habitat unknown.

P. funebris RvE., Conch. Icon., f. 54. Dec., 1854.

P. STELLÆFORMIS Reeve. Pl. 17, figs. 25, 26, 27; pl. 61, figs. 62–65.

Shell solid, low-conic, angularly oval, the apex central. Surface vary irregularly and roughly sculptured with carinated radiating ribs and riblets, 8 or 9 being more prominent in the typical form of the species. The ribs are rude, irregular, often somewhat scaly. White, sometimes marked in the interstices with black or rusty-black.

Interior white, frequently slightly stained in places with yellowish; central area white or stained with fleshy-brown, the musclescar sometimes outlined with reddish-brown. Margin very irregularly toothed. Length 45, breadth 35, alt. 14 mill.

Japan to Port Jackson, S. Australia; eastward to Viti, Cook's and Society Archipelagos.

P. stellæformis REEVE, Conch. Syst. ii, p. 15, t. 136, f. 3, 1842.—DUNKER, Ind. Moll. Mar. Jap., p. 156.—*P. pentagona* REEVE, Conch. Icon., f. 48, 1854—LISCHKE, Jap. Meeresconchyl. i, p. 114 (not *P. pentagona* Born).—ANGAS, P. Z. S. 1867, p. 221.—DALL, Amer. Journ. Conch. vi, p. 272, t. 15, f. 22 (dentition).—*P. paumotensis* GLD., Proc. Bost. Soc. N. H. ii, p. 150, 1846; U. S. Expl. Exped. Moll. & Sh., p. 339, f. 440.—*P. cretacea* REEVE, Conch. Icon., f. 53.—*P. tramoserica* AD., Ann. Mag. N. H. 1868, p. 369.

This excessively variable species is allied to *P. pica,* differing mainly in not being narrowed anteriorly as a general rule. The main distinction, however, is geographic, the present form being Pacific, in distribution, whilst *pica* (and its immediate allies or varieties) is from the western part of the Indian Ocean. I doubt the occurrence of true *stellæformis* in the last mentioned area.

It is impossible to say what *P. pentagona* Born (Mus. Test. Cæs. Vindob., p. 421, t. 18, f. 4, 5) is intended for. It is certainly not this species. Von Martens surmises that it may not belong to the *Patellidæ.* The figures somewhat resemble a large *Siphonaria.*

The typical form of *stellæformis* has 8 to 10 larger ribs.

Among the large number of minor modifications, typically quite diverse but intergrading by easy stages with the types, the following may be noticed:

Var. PAUMOTENSIS Gld. (pl. 47, figs. 4, 5). Outline much more regularly oval; riblets very numerous and subequal. *P. cretacea* Rv. is a synonym.

Another slightly differing form is figures on pl. 61, figs. 62, 63, 64. It is large, oval, coarsely ribbed, with fine secondary radiating striæ. The interior is marked with brown.

The specimen figured on pl. 61, fig. 65 has a great similarity to Reeve's *P. stellaris,* and I am inclined to believe it is the same. It is very distinctly octoradiate, the ribs wide; both ribs and intervals finely striated radially. See p. 51 of this volume for remarks on Reeve's *stellaris,* and pl. 36, figs. 65, 66, copies of the original figures.

Var. NIGROSULCATA Reeve (pl. 61, figs. 66, 67). "Ovate, rather solid, radiately grooved, grooves narrow, rather distant; rough chalk-white, grooves more or less black; interior yellowish-white, border faintly lineated." This seems to be a regularly oval form smaller than v. *paumotensis*, and more stained with rust-red inside. Numbers of shells before me correspond with Reeve's figures.

P. ACULEATA Reeve. Pl. 25, figs. 20, 21 ; pl. 62, figs. 71, 72, 73.

Shell oblong-oval, solid, conical, the *apex at the front third*. Surface dull, having numerous (about 23) strong, carinated, and more or less scaly-nodose ribs. White, with inconspicuous rust-reddish concentric bands.

Interior whitish, tinged toward the middle with orange-brown, the cavity of the apex white with blue-black stains. Edge of shells dentate. having several narrow short reddish lines in each interval.

Clarence River to Twofold Bay, Australia ; Tasmania.

P. aculeata REEVE, Conch. Icon., f. 90, 1855.—ANGAS, P. Z. S. 1867, p. 221.—TENSION-WOODS, Proc. Roy. Soc. Tasm. 1877, p. 22. —BRAZIER, Proc. Linn. Soc. N. S. W. xiii, p. 224, 1883.—*P. squamifera* Rv., *l. c.,* f. 94.—ANGAS, *l. c.,* p. 221.

There are about 23 large ribs, and some smaller ones in the interstices. The apex is at the *anterior third*, not *central* as it is in *P. stellæformis*, and the ribs are scaly. It is a common form at Port Jackson. The description of Reeve's *P. squamifera*, which Brazier considers a mere synonym of this variable shell, is as follows :

P. squamifera Reeve (pl. 62, figs. 74, 75). Shell ovate, somewhat depressed, rather thick, apex nearly central ; radiately roughly ribbed and ridged, ridges irregularly rudely scaled. Whitish tinged with ash and black. Interior bluish-white.

This is a solid, ash-colored shell, roughly sculptured throughout with irregular scaly ribs and ridges. (*Rve.*)

P. MORBIDA Reeve. Pl. 15, figs. 3, 4.

Shell ovate, rudely depressed, rotundately raised in the middle, radiately fimbriately ridged, more or less eroded, ridges obsoletely short-spined. Interior yellowish-white, more or less irregularly stained with black ; exterior rust-eroded. (*Rve.*)

Cape of Good Hope.

P. morbida RVE., Conch. Icon. f. 64, Jan. 1855.

Of a characteristic depressedly furbelowed growth round the margin, the radiating ridges armed here and there with short, sharp black and white spines. Exteriorly the shell is roughly rust-eroded; interiorly it has a peculiarly diseased look. (*Rve.*)

P. CHAPMANI Tenison-Woods. *Unfigured.*

Shell ovate, somewhat broad behind, reddish or scorched and nebulously brown, apex acute, submedian; with 8 radiating ribs more or less valid, and depressedly rounded, profusely radiate with very fine liræ, and girdled with irregular sulci; margin angulate, nodulose. White within and clouded pale rose color, spatula scarcely visible. Length 20, breadth 15, alt. 5 mill. (*T.-W.*)

Tasmania.

P. chapmani T.-W., Proc. Roy. Soc. Tasm. for 1875, p. 157, 1876.

Very rare. Four of the ribs are posterior, and the four anterior are smaller. (*T.-W.*)

P. USTULATA Reeve. Pl. 22, figs. 11, 12.

Shell somewhat squarely ovate, a little attenuated in front, posteriorly convexly depressed, anteriorly tumidly umbonated, apex obtuse; radiately elevately striated, striæ scabrous next the margin; burnt-red, neatly rayed with rather distant narrow white bands, striæ more or less black next the margin, interior white. (*Rve.*)

Tasmania (Tenison-Woods).

P. ustulata RVE., Conch. Icon f. 88, Jan., 1855.—TENISON-WOODS, Proc. Roy. Soc. Tasm, for 1876, p. 49, 1877.—*P. tasmanica* T.-W., *l. c.* for 1875, p. 157, 1876.

Tenison-Wood's description of his *P. tasmanica* is as follows:

Shell ovate, solid, sordidly yellowish white, often corroded, apex sub-median with about 21 valid, angular radiating ribs, and the interstices rayed profusely with very fine subimbricated liræ; within ivory white and shiny, more or less tinged with yellow; margin narrow, elegantly pectinated; margined with a very fine blue line within, and an interrupted dusky brown line outside. Spathula scarcely defined.

Recherche Bay and south generally. Nearer to *P. alticostata Angas* than any other.

Length 49, width 38, alt. 20 mill. (*T.-W*).

"If I am right in my identification of this shell, it must be the same as my *P. tasmanica* described in last year's Proceedings of this Society. Reeve gives no habitat for his shell, which from appearance was worn and corroded. The unworn specimens found living

on the rocks are as different as possible, the ribs and riblets being
then conspicuous, and the whole shell a dull yellowish-white with
no trace of the scorched coloring.   When dead, however, and
thrown on the beach this feature is conspicuous.   It has many fine
riblets between the coarse, somewhat nodular ribs, and the margin
is very finely pectinated.   A peculiarity of the animal is that it
seldom comes above low water mark, and prefers situations where
it is much exposed to the waves.   It is very stationary, often being
sunk into a regular pit in the rocks, and appears to live upon the
fine green ulva on the rocks.   It is nearly always covered, not
only with confervoid growths, but also nulliporæ so as to quite
alter its shape and appearance.   This often alters the height of the
shell, which is usually depressed, and changes the position of the
apex, which is usually submarginal.   The interior is white and the
spatula not defined.

"The animal is of uniform pale yellow at the base; white above
the foot, gills semi-pellucid and continued as a delicate fringe *all
round* the mantle   I, however, noticed one exception where, like
the former species, the gills were discontinued in front of the head,
mantle without tentacles ;   head livid, with semi-pellucid tentacles ;
eyes very small and at exterior base ;   buccal mass red and fleshy ;
cartilaginous jaws long and less tumid than most limpets ;   odontophore *scarcely as long as shell ;*   not coiled, but bending with intes-
tine in two folds.   Teeth closely set and not high, composed of five
central small curved cusps, and two tri-lobed laterals, all narrowly
tongue-shaped, laterals more acute.   The five centrals have the
middle tooth often small.   Teeth brown, lighter on the summit."

(3) GROUP OF P. GRANULARIS.

Shell oval, sculptured with numerous granose riblets, none of them
notably larger.   Central tract of the inside and border generally
dark. Dist., S. Africa.

P. GRANULARIS Linné.   Pl. 63, figs. 80, 81, 82, 83.

Shell solid, conical, ovate; apex in front of the middle; sculptured with numerous (about 50) regularly *granose riblets ;* the granules usually like small solid scales.   Color dull brown, blackish, or
ashen, dull reddish above.

Inside *opaque-white, with a broad black or dark brown border* and
a large *reddish-chestnut central area.*

Length 58, breadth 48, alt. 26 mill.

Cape Good Hope.

P. granularis Linn., Syst. Nat. x, p. 782.—Hanley, Sh. of Linn. p. 419.—Lam., An. s. Vert. vi, p. 330.—Reeve, Conch. Icon. f. 31.—Krauss, Die Südafric. Moll. p. 52.—Quoy & Gaimard, Voy. de l'Astrol. iii, p. 341, t. 70, f. 12–15.

Characterized by the closely granose riblets of the outside and the broad dark border and large orange-brown central area of the interior. It is not at all iridescent within.

In one specimen (pl. 63, fig. 83) of this species before me the earlier portion is black, finely but obsoletely radiately striated, differing totally from the normal later growth. A similar change of structure has already been noticed in this volume, page 18, under *Acmæa pelta* var. *nacelloides.*

P. NATALENSIS Krauss. Pl. 13, figs. 65, 66, 67.

Shell oval or oblong, conical, rather solid, slopes straight; apex at the front third. Surface sculptured with 40–46 rather separated regularly granulose riblets. Color ashen or blackish, the granules black; apex generally eroded.

Interior white, the margin intensely black, having a border of blackish-brown flames; central area dark reddish-brown.

Length 31, breadth 23, alt. 12 mill.
Length 27, breadth 20, alt. 11 mill.

Natal, northward to Guinea.

P. natalensis Krauss, Die Südafric. Moll., p. 53, t. 3, f. 10, 1848. —*P. echinulata* Krauss, *l. c.,* p. 52, t. 3, f. 15.—*P. nigrosquamosa* var. *miliaris* Phil., Zeitschr. f. Mal. 1848, p. 162.—*P. nigrosquamosa* var. 1, Dkr., Ind. Moll. Guin. Infer., p. 41, t. 7, f. 7, 8.

This is a compact, long-oval species. The outside is blackish with black tubercles on the riblets. The eroded apex is brown or clouded with brown, and surrounded by a white tract, which usually is digitate or rayed more or less, as figure 66 shows. The black outer layer is quite thin. The margin is intensely black. The central spatula varies from olive-brown to a deep red-brown. Philippi's *miliaris*=Dunker's var. 1 of *nigrosquamosa* (pl. 13, figs. 63, 64) is a synonym.

Var. ECHINULATA Krauss. Pl. 13, figs. 60, 61, 62.

Smaller, narrower; grayish-brown, rayed with whitish. The basal side-margins are somewhat raised, so that the shell rests upon the ends only.

Table Bay.

P. VIDUA Reeve.   Pl. 63, figs. 78, 79.

Shell ovate, moderately convex; strongly, sharply ribbed, ribs alternately smaller, minutely scaled, scales distant, the alternate rib sometimes nearly obsolete; rusty-black, eroded at the apex. Interior whitish, with a broad rusty-black band at the edge; nucleus pale rust. (*Rve.*)

Island of Camiguing, Philippines.

P. vidua RVE., Conch. Icon. f. 22, Oct. 1854.

A moderately convex shell, rayed with sharp ribs, alternately larger and smaller, each rib being roughened with minute, somewhat distant scales. The chief characteristic of this species lies in its broad deep rust-black marginal border. (*Rve.*)

(4.) GROUP OF P. COCHLEAR.

Shell spoon-shaped, produced and narrowed in front; ribs numerous, subequal.

P. COCHLEAR Born.   Pl. 27, figs. 34, 35.

Shell *spoon-shaped*, depressed, solid; apex subcentral. Surface having numerous close radiating riblets, grayish or blackish, usually encrusted or eroded.

Interior white or purplish-blue, the *muscle-scar black.*

Length 60, breadth 45, alt. 15 mill.

Cape Good Hope.

P. cochlear BORN, Mus. Cæs. Vindob. p. 420, t. 18, f. 3.—REEVE, Conch. Syst. ii, t. 136, f. 5; Conch. Icon. f. 24.—KRAUSS, Die Südafric. Moll. p. 48.

The curiously narrowed anterior end gives a spoonlike appearance to this shell. Some specimens are almost perfectly flat, and have a red central callus.

It has been made the type of a subgenus by the Adams brothers, but a number of other species approach it in contour, and form connecting links with the oval limpets.

(5) GROUP OF P. LONGICOSTA.

Shell large, having some (usually 7–11) of the ribs decidedly larger, rendering the outline more or less polygonal.   Distribution, S. Africa.

P. PATRIARCHA Pilsbry. Pl. 64, figs. 84, 85 ; pl. 65, fig. 86.

Shell *very large and solid*, conical, the apex a little in front of the middle; rounded oval, nearly as wide as long. Surface dull, coarsely ribbed, 9 or 10 primary ribs radiating from the summit, the secondary ribs numerous, some of them nearly as prominent as the primaries ; radiating striæ also are to be seen in some places ; growth-lines fine, inconspicuous. Color dull reddish-ashen.

Interior *pure white*, the edge reddish-brown, bordered by a narrow subtranslucent band, sometimes not conspicuous.

Length 127, breadth 112, alt. 45 mill.

Length 117, breadth 105, alt. 43 mill.

Cape Good Hope.

? *P. rustica* REEVE, Conch. Icon., f. 8, 1854, not *P. rustica* Linné (1758), nor *P. rustica* Menke (1843.)

This is, next to *P. (Ancistromesus) mexicana*, the largest limpet I have seen, one of the specimens before me attaining a length of five inches. The interior is pure white, *totally lacking the fibrous texture* which renders *P. cærulea* and its allies iridescent. The shell figured by Reeve as " *P. rustica* Linn." is probably an immature specimen of this species, having the sculpture sharper. The true *rustica* of Linné is the small shell known to us as *P. lusitanica* (*q. v.*). The *rustica* of Menke is the same as *P. neglecta* Gray (*q. v.*)

P. TABULARIS Krauss. Pl. 16, figs. 9, 10.

Shell ovate, solid, much depressed, whitish, with radiating reddish lines ; radiately ribbed and striate, the striæ and ribs unequal, carinated, *scaly;* larger ribs 12–14 in number. Margin digitately toothed ; vertex obtuse, situated at two-fifths to one-third the length.

Inside white, having a peripheral border of dull yellow spotted with brown, 2 mill. wide ; space between border and muscle-impression white ; the muscle-scar and central area pale brownish.

Length 59, breadth 46, alt. 8 mill.

Table Bay, S. Africa.

P. tabularis KRAUSS, Die Südafric. Moll. p. 47, t. 3, f. 8.—DKR., Ind. Moll. Guin. Infer. p. 41.—*P. obtecta* KR., *l. c.* p. 49, t. 3, f. 11.

Reported by Dunker from Benguela, W. Africa. The strongly angular, star-like form and closely scaly ribs and striæ are the prominent characters of this shell.

Var. OBTECTA Krauss.   Pl. 16, figs. 7, 8.

In sculpture like *P. tabularis;* but narrower, more elevated, less strongly ribbed.   The central area of the interior is brown.

Length 30, breadth 22, alt. 8 mill.

Table Bay.

P. GRANATINA Linné.   Pl. 62, figs. 76, 77.

Shell large, rather thin but solid, conical, angularly ovate; apex subcentral, eroded, rust-color or dark brown.   Surface having unequal radiating ribs or carinæ, of which one at each side and three in the rear are especially prominent.   Color dull whitish, *closely marked with black spots,* often zigzag or angular.

Interior *white,* the portion outside of the muscle-scar having a distinctly *fibrous* appearance; *central area blackish-brown,* sometimes mottled with white.   *Border very narrow,* finely dotted with black and brown.   Length 85, breadth 72, alt. 30 mill.

Cape of Good Hope.

P. granatina L., Syst. Nat. x, p. 782.—LAMARCK, An. s. Vert. vi, p. 324.—KRAUSS, Die Südaf. Moll., p. 43.—REEVE, Conch. Icon., f. 4.—*P. apicina* LAMARCK.—DELESSERT, Rec. de Coq., t. 21, f. 4.— ? *P. picta* PERRY, Conchology, t. 43, f. 7.

The prominent ribs of the exterior and the spotting of black (sometimes lost by erosion) are characteristic; inside the white, fibrous appearance and deep brown center are excellent diagnostic points.

In old specimens the dark center is often considerably invaded by white in the middle and forward.

P. OCULUS (Born) Auct.   Pl. 27, figs. 30, 31, 32.

Shell large, angularly oval, conic or depressed, solid.   Apex in front of the middle.   Sculptured with large angular unequal ribs, which project at the margins; and having a secondary sculpture of radiating striæ when not eroded.   Color blackish or dull brown outside; usually eroded.

Interior having a *very broad blackish-brown border,* a light zone just outside the muscle-impression, the latter strongly marked, bluish-white.   *Area inside the impression callously thickened, yellowish flesh-colored.*

Length 88, breadth, 75, alt. 25 mill.

Length 64, breadth 63, alt. 12 mill. (younger shell).

Length 110, breadth 106, alt. 42 mill. (largest specimen seen).

Cape of Good Hope.

P. oculus BORN (in part), Mus. Cæs. Vindob. p. 418.—SOWB., Conchol. Manual, f. 229.—REEVE, Conch. Icon. f. 2.—*P. badia* Gmel. Syst. p. 3700.—*P. fuscescens* GMEL., Syst. p. 3701.—*P. schrœteri* KRAUSS, Die Südafric. Moll. p. 43.

Allied to *P. granatina*, but not variegated outside except in the young (fig. 32), and entirely different in color inside ; this species being broadly black bordered with a light central area. The more markedly stellate forms, such as fig. 31, approach *P. longicosta* in outline, but the coloring of the interior is constantly distinct.

It seems advisable to retain the well-known name *oculus* for this species, despite the fact that Born included another and probably distinct species with this, in his references.

P. LONGICOSTA Lamarck. Pl. 28, figs. 37, 38.

Shell depressed, *star-shaped*, having 7 to 9 principal ribs which are *carinated and project at the margins*, and a variable number of smaller projecting ribs. Color black, usually rayed with whitish.

Interior white, the central area often yellowish or flesh-colored, in the young generally stained or mottled with blue-black. The laciniate edge is bordered with black, often dotted with gray-white, rarely entirely white. Length 75, breadth 70, alt. 20 mill.

Table Bay, Cape of Good Hope.

P. *longicosta* LAM., An. s. Vert. vi, p. 326, 1819.—DELESSERT, Rec., t. 21, f. 3.—REEVE, Conch. Syst. ii, p. 136, f. 6 ; Conch. Icon., f. 11.—POT. & MICH., Galerie, i, t. 37, f. 7, 8.—*P. gorgonica* HUMPH. mss., *teste* Reeve.

Allied to some forms of *P. oculus*, but much more distinctly stellate, and having a narrower black border within.

Section *Ancistromesus* Dall, 1871.

Ancistromesus DALL, Amer. Jour. Conch. Apr. 4, 1871, p. 276.

Shell very large and heavy, *its inner layer porcellanous, opaque*.

Animal with a complete branchial cordon, the lamellæ long and slender, subequal ; sides of foot smooth ; radula furnished with a simple rhachidian tooth having a cusp ; the two inner laterals on each side anterior to the third pair, which are large and quadridentate ; uncini simple.

Formula of teeth $_3(_1{}^{2\cdot2}{}_1)_3$. Pl. 31, fig. 62.

The shell resembles very closely, in texture, sculpture and form, the larger South African species of Patella (*Scutellastra* of my arrangement), some of which also possess a rhachidian tooth, as Hogg and others have demonstrated.

P. MEXICANA Broderip & Sowerby. Pl. 31, figs. 59–62.

The shell is very large, thick and heavy, oval, conical, with central summit. The dull, soiled white, eroded surface shows about 10 low angles or obsolete ridges, and young shells are rather finely striated (fig. 61).

The interior is pure white or tinged with flesh-color, having also, usually, some brown or purplish stains. The muscle-scar is conspicuous, roughened.

Length 200, breadth 150, alt. 80 mill.

Mazatlan! San Blas! Acapulco! W. Mexico; also Central America.

P. mexicana B. & S., Zool. Journ. iv, p. 369.—MKE., Zeitschr. f. Mal. 1851, p. 37.—REEVE, Conch. Icon. f. 1.—CARPENTER, Mazat. Cat. p. 199.—*Ancistromesus mexicanus* DALL, Amer. Journ. Conch. vi, p. 266, t. 15, f. 21 (dentition).—*Lottia gigantea* GLD., ms. in B. M.—*Patella maxima* ORB., Voy. Amér. Mérid. p. 482, and in B. M. Catal. d'Orb. Moll., p. 53.

This is the largest living species of limpet, frequently attaining a length of 8 to 14 inches. The animal is black, more or less marbled and streaked with white. The shell is often used as a wash-basin in Central America. (See Dall, *l. c.*)

D'Orbigny described this species as *P. maxima*, giving the locality Payta, Peru.

Subgenus HELCION Montfort, 1810.

Helcion MONTF., Conch. System. ii, p. 62.—GRAY, Guide, p. 176. —DALL, Amer. Journ. Conch. vi, p. 276.

Helcion is composed of limpets differing from *Patella s. s.* in having the gill-cordon interrupted in front, the shell cap-shaped, apex curving forward. Two sections are distinguishable: HELCION (restricted), having a strongly convex solid shell with scaly radiating ribs, and PATINA, in which the shell is nearly smooth and thinner.

Section HELCION s. s.

The gill-cordon is interrupted over the head, composed of small and filiform strands.

The dentition is said by Dall to be the same as *Patella* except that the third or outer cusp of the third lateral tooth is obsolete.

The shell is cap-shaped, scaly-ribbed, the apex strongly curving forward.

P. PECTINATA Born. Pl. 51, figs. 1, 2, 3.

Shell solid, oval, elevated, cap-shaped, *the apex curved forward*, nearly to or over the anterior margin. Surface sculptured with numerous close, densely prickle-scaled riblets, alternately larger and smaller. Riblets black, the intervals buff or pinkish.

Interior of a dull lead color, sometimes a little iridescent at the edge. Length 27, breadth 22, alt. 14 mill.

Cape of Good Hope.

Patella pectinata BORN, Mus. Cæs. Vindob. p. 423, t. 18, f. 7, (1780).—LAM., An. s. Vert. vi, p. 334.—POT. & MICH., Galerie, i, p. 529, t. 37, f. 11, 12.—KRAUSS, Die Südafric Moll. p. 57.—Not *Patella pectinata* LINN., Syst. Nat. x, p. 783, nor of GMELIN, Syst. xiii, = *Siphonaria!*—*Patella intorta* SOWERBY, Genera of Shells, Cephala, *Patella*, f. 5.—*Patella pectunculus* GMELIN, Syst. xiii, p. 3713.—*Helcion pectinatus* MONTF., Conch. Syst. ii, p. 63.—GRAY, Guide Syst. Dist. Moll. B. M., p. 126 (descr. of branchiæ).—DALL, Proc. Acad. N. S. Phila. 1876, p. 244 (dentition).

This cap-shaped, scale-ridged black species is unlike any other limpet, having much the contour of *Scutellina*. Linnæus is generally but erroneously quoted as the authority for the name. It is a common Cape species.

Section PATINA (Leach) Gray, 1852.

Patina LEACH, Moll. Gt. Brit. (Gray's edition) p. 223, 1852.—GRAY, Syn. Br. Mus. 1840 (name only no definition; no type mentioned).—*Ansates* SOWB., Conch. Man. edit. ii, p. 68.—*Helcion, Patella* and *Nacella* sp., of authors.

The branchial cordon is interrupted in front; side of the foot without an epipodial ridge or papillæ.

Dentition, pl. 52, fig. 2. Two inner lateral teeth on each side anterior, having simple cusps, the third lateral having a broad tripartite cusp. Formula of dentition $_3(_1^{202}{}_1)_3$.

Epipodial papillæ have been ascribed to this group by a recent authority, but I have satisfied myself that none are present by an examination of specimens. The sides of the foot are as smooth as in *Helcioniscus exaratus*.

The shell has the contour and texture of *Helcion* but is generally thinner and the radiating sculpture is obsolete.

P. PELLUCIDA Linné.   Pl. 51, figs. 4, 5, 9, 10.

Shell thin, oval, elevated, the apex curved forward, near the anterior end.   Surface polished, smooth except for very faint radiating striæ.   Dark olive or brownish horn-color, reddish or blackish at the apex, and usually having a few radiating interrupted lines of vivid blue.   Interior brownish, reddish or dusky within the cavity. Length 20, breadth 15, alt. 8 mill.

> *Lofoten, Norway, to Cascaes Bay, Portugal.*

P. pellucida L., Syst. Nat. x, p. 783.—FORBES & HANLEY, Hist. Brit. Moll. ii, p. 429, t. 61, f. 34; t. AA, f. 1 (animal).—*P. cærulea* PULT., Cat. Dorset., t. 23, f. 6.—*P. bimaculata* MONT., Test. Brit., p. 482, t. 13, f. 8.—*P. cæruleata* DA COSTA, Brit. Conch., p. 7, t. 1, f. 5, 6.—*P. elongata* and *elliptica* FLEM., Encyc. Edin., t. 204, f. 2, 3.— *P. cornea* POT. & MICH., Galerie Douai, p. 525, t. 37, f. 5, 6.—? *P. intorta* PENNANT.—*P. minor* WALLACE and *P. cornea* MICHAUD, *teste* Jeffr.—*Patina pellucida* LEACH, Moll. Gt. Brit., p. 224.—DALL, Amer. Journ. Conch. vi, p. 280, t. 16, f. 30 (dentition).—*Helcion pellucidum* JEFFR., Brit. Conch. iii, p. 242, t. 5, f. 4.—*Nacella pellucida* SARS, Moll. Reg. Arct. Norv., p. 119, t. 2, f. 8 (dentition). —*P. cypridium* PERRY, Conchology, t. 43, f. 6.—*P. lævis* PENNANT, Brit. Zool. iv, p. 144, t. 90, f. 151.—*Patina lævis* LEACH, *l. c.*, p. 224. —*Patella cornea* HEEBLING, Beiträge zur Kenntniss neuer u. seltener Conchyl., in Abhandlungen einer Privatgesellschaft in Böhmen zur Aufnahme der Mathematik, der Vaterländischen Geschichte und der Naturgeschichte, iv, p. 107, t. 1, f. 8, 1779.

A delicate cap-shaped shell, common on fronds of laminariæ throughout the seas of northern Europe.

Var. LÆVIS Pennant.   Pl. 51, figs. 6, 7, 8.

Shell more erect, the summit more nearly central; solid, thick, more obviously radiately striated and having coarse concentric wrinkles.   Length 22, breadth 18–20, alt. 8–10 mill.

The distribution is the same as the typical *pellucida*.   The differences are the result of station, the *lævis* living imbedded in the stems of fuci.   Specimens of all sizes may be found having the typical *lævis* form, but frequently a *lævis* is surmounted by an earlier growth of the *pellucida* type, like the similarly caused forms of *P. granularis, compressa, Acmæa pelta,* etc., etc.

P. TELLA Bergh. Pl. 51, figs. 12–26.

Under this name Bergh has given the following description and figures of the soft parts of a specimen, the shell of which had been detached and lost.

The body measured in length 8·5, breadth 5 mill. The sole of the foot (fig. 14) is oval, 8 mill. long, 4½ broad. The color is light brown, the sole having a median longitudinal band shining like a tendon, not quite reaching to the posterior end. The foot was very strong, having a narrow fringe except at the head and behind, but not scalloped as it is in *P. pellucida*. The branchial cordon is interrupted in front. The head is strong, exactly similar to that of *P. pellucida*. At the three-cornered, kidney-shaped anterior end (fig. 13) is the broad three-cornered mouth, and behind it the oblique, as if cleft, front end of the buccal mass. The tentacles were pretty short, quite cylindrical (figs. 12, 13), similar to those of *P. pellucida*. Unlike the latter, the eyes were not visible through the integument. On the upper side of the head the radula showed blackly through. The mantle-margin shows none or slight trace of a clothing with closely placed, short, tentacular bodies. The positions of anal and infra-anal papillæ could not be determined.

Above in the mouth-opening projected the edge of the upper jaw. The buccal mass was strong; about 2.5 mill. long by 1.5 broad. The form of the basal-plate of the jaw was not observed. The cutting (anterior) plate (fig. 17) was large, 1·3 mill. broad, light brownish-yellow, darker at the back margin, half-moon-shaped, a little narrower in the middle than at the sides, with obliquely excavated anterior margin, thin back margin. The tongue was similar to that of *P. pellucida*, strongly black-pigmented at the side areas. The teeth had fallen off. The number of rows, however, seems to have been 11. The odontophore sheath is very long, reaching over the upper surface of the foot. The posterior end is lacking; the remainder has a length of 7 mill., its middle brownish, the sides with a peculiar greenish-yellow luster. In the sheath there are 38 developed, 2 nearly developed (lighter colored) and 6 colorless, undeveloped, transverse series of tooth-plates. The dentition agrees in all important characters with that of *P. pellucida*, as figured by Lovén. In each row there are 12 teeth, the formula being $3(\text{x}_4^{202}\text{x})_3$. On the rhachis there is in the middle-line (figs. 19, 20), a low, elongated, narrow (median) ridge, without cusp (homologous with the rhachidian tooth in *Ancistromesus*), and on each side of it three strong lateral teeth, of which the inner one is more or less fused with the median at their bases. The inner lateral (fig. 21) and the median lateral were rather similar to one another, the second lateral (fig. 19a, a, 20a, a) being only somewhat stronger, with somewhat S-shaped base (fig. 19); the cusps being on both strong but narrow. The outer lateral was much stronger than the others; the basal-plate broader (fig 25); the cusp broad, three-cuspidate, the outer cusp shovel-shaped, the two inner narrower, more acute. Of the 3 uncinal teeth on the *pleura* of each side (figs. 23–26), the outer is larger (fig. 23), the inner the smallest. They show themselves often as if fastened on the upper part of the basal-plate of the outer lateral tooth (figs. 23, 24). All three pairs of lateral teeth show the basal part and the outer portion of the cusp amber-yellow, the intermediate part was almost as clear as glass. The side cusps were light horn-yellow. The length of the basal-part of the middle lateral tooth is about 0·15 mill.; the height from 0·15–0·16 mill. The length of the base of the outer lateral tooth 0·145–0·15, the breadth 0·14 mill.; the height of the tooth 0·16–0·18 mill.; the breadth of the cusp was about 0·12–0·13 mill.; the length of the outer side-cusp 0·1, the middle 0·08, the inner 0·056 mill. The cusp of the outer was 0·025 mill. high.

In the stomach was found a large-celled vegetable mass, similar to that found in various *Pleurophyllids*.

Sargasso Sea.

Patina tella BERGH, Beiträge zur Mollusken des Sargassomeeres, in Verhandlungen der k.-k. zool.-bot. Gesellschaft in Wien, xxi, p. 1297, t. 12, f. 12–26 (1871.)

The explanation of figures is as follows:

12. Head, from above.

13. Head, from in front.

14. Sole of the foot.

15. Buccal mass from the side; *a* anterior end, *b* broken radula sheath.

16. Buccal mass from below, *a*, *b*, as above.

17. Front or cutting plate of the jaw, cutting edged turned upward.

19. Median ridge, inner and middle lateral teeth from the under side.

20. Median ridge, inner and *a* middle lateral tooth, obliquely from beneath.

21. Inner lateral tooth, from the side.
23. Outer lateral tooth and side cusp, from the inner side.
24. The same, from the outer side.
25. The same, from the posterior side.
26. The three uncini.

P. PRUINOSA Krauss. Pl. 51, figs. 11, 11 ; pl. 13, figs. 68, 69.

Shell oval, depressed-conical, nearly smooth, the apex near the front fourth of the length. Surface having faint radiating striæ. Color varying from yellowish-olive to blackish-olive often mottled or rayed, and having fine interrupted radiating lines of blue. Interior olivaceous, dusky-whitish in the cavity.

Length 28, breadth 21, alt. 6 mill.
Length 31, breadth 24, alt. 9 mill.

Cape Good Hope.

P. pruinosa KRAUSS, Die Südafric. Moll., p. 56, t. 3, f. 9.—REEVE, Conch. Icon., f. 109.—*Patinastra pruinosa* THIELE, Das Gebiss der Schn. ii, p. 326, t. 28, f. 24, 25 (dentition and jaw.)

It is larger and more depressed than *P. pellucida.* The blue lines are broken into minute dots. It has the same indistinct radiating striæ that are to be seen on the European species.

Dr. Thiele has made this the type of a new genus, *Patinastra*, founded upon a very slight difference in the dentition, which is intermediate between that of *Patina* and *Helcion.*

* * *

P. ROSEA Dall. Pl. 50, fig. 44.

Shell small, egg-ovate, of a deep rose color ; externally smooth except for very faint radiating ridges divaricating from the apex, and for lines of growth. Margin entire ; apex minute, produced before the anterior margin. Interior smooth, white except the margins, which are polished and of the same color as the exterior. Nacre, especially when weathered, silvery. Length ·35, width ·27, alt. ·12 inch, of largest specimen. (*Dall.*)

East side of Simeonoff Island, Shumagins.

Nacella (?) rosea DALL, Proc. Cal. Acad, Sci. iv, p. 270, t. 1, f. 2 (Oct. 8, 1872.)

The soft parts have not been examined. The position of this shell in *Nacella* where originally placed, is therefore, doubtful. It may
8

prove to be a *Patina*, which the types, seen by me in the U. S. National Museum, resemble as much as they do *Nacella*.

Subfamily NACELLINÆ, Thiele.

The researches of Thiele have demonstrated that there are but *two* lateral teeth on each side in *Nacella*, *Patinella* and *Helcioniscus*, whilst *Patella*, *Helcion*, *Patina*, etc., possess *three* on each side. This difference is undoubtedy of considerable value, and I therefore depart from the arrangement adopted in the synopsis of groups on page 79, and consider the forms in my second division "*B. One inner lateral tooth on each side anterior*" as constituting the subfamily NACELLINÆ.

The subfamily differs from *Patellinæ* in possessing the characteristic dental formula $_3(_1^{1\cdot1}{}_1)_3$, and in the shells having a distinctly metallic luster inside ;—the genus *Patella* having the formula of teeth $_3(_1^{2\cdot12}{}_1)_3$ and the inside of the shell either transparent and fibrous, or opaque, porcellanous.

On pages 79, 80, I have divided this group into two subgenera : (I) NACELLA, with sections *Nacella* s. s. and *Patinella* Dall, and (II) HELCIONISCUS Dall. These divisions are used in the same limits by Thiele (*l. c.*), except that he considers the three as of generic rank, as Dall has already done.

Genus NACELLA Schumacher, 1817.

Nacella SCHUM., Essai d'un nouv. Syst., p. 179.—DALL, Amer. Journ. Conch. vi, p. 274, 1871.—THIELE, Das Gebiss der Schn. ii, p. 329.—not *Nacella* of CARPENTER, SARS, *et al.*

The gill-cordon is continuous.

The foot is encircled by a scalloped epipodial ridge, interrupted in front.

The dentition is practically the same as in *Helcioniscus* (*q. v.*), differing notably from that of *Patella* and *Helcion*+*Patina*.

The shell has the apex subcentral or anterior, and is characterized by a peculiarly metallic texture, having the central area of the interior generally of a red-bronze color.

The shell and dentition of *Nacella* approach near to *Helcioniscus*, but from this and all other *Patellidæ* it is sundered by the presence of a developed epipodial ridge.

Cape Horn was evidently the birth place of *Nacella* and *Patinella*. Thence they have been distributed eastward to the Falkland, New

Georgia and Kerguelen islands, by the eastward sweeping Antarctic current, carrying them upon sea weeds.

A number of West American species placed under *Nacella* by Carpenter and others belong to *Acmæa*.

Two sections, having but slight distinctive characters in either shell or soft parts, have been instituted:

(1). *Nacella s. str.*, having the gills smaller in front, the shell thin, light, nearly smooth, the apex far forward.

(2). *Patinella* Dall, in which the gills are equal all around, the shell more solid, and deeply colored.

Section *Nacella* Schum. (restricted.)

The dentition has been figured by Dall, and lately with more exactness by Thiele.

According to Thiele, the typical *N. mytilina* is more closely allied in dentition to *Patinella* than to the other species of *Nacella* recognized by him, *N. vitrea* and *N. hyalina*. The rhachidian tooth is rudimentary and bears no cusp. The inner side-tooth (homologous, according to Thiele, with the second lateral in *Patella*) is broad, truncated, and has an outer cusp; the second side tooth is posterior, and bears a small cusp on each side of the broad, truncated median cutting edge. Pl. 74, fig. 3, represents *N. vitrea* Phil.; pl. 74, fig. 4, represents *N. mytilina* Helbl.

Only one species of this section, as restricted, is known. It lives upon the great sea weeds of the Tierra del Fuego shores.

N. MYTILINA Helbling. Pl. 50, figs. 32–39.

Shell thin or fragile, elliptical, convex, the apex strongly curved forward and downward near the anterior end; surface smooth except for narrow, faint, separated radiating riblets, more distinctly developed in front. Color usually light brownish, the apex coppery.

Inside silvery, large specimens usually having a coppery stain in the middle. Length 41, breadth 24, alt. 17 mill.

Sts. of Magellan; Kerguelen Is.

P. mytilina HELBLING, Abhandl. einer Privatgesellsch. in Böhmen zur Aufnahme der Mathematik, der vaterländischen Geschichte u. der Naturgeschichte, iv, p. 104, t. 1, f. 5, 6, (1779).—*P. mytilina* GM., Syst. xiii, p. 3698.—SCHUB. & WAGN., Conch: Cab. xii, p. 124, t. 229, f. 4052, 4053.—*P. conchacea* GM., *l. c.*, p.

3708.—*Nacella mytiloides* Schum., Syst. Vers Test. p. 179.—*N. cymbularia* Lam., An. s. Vert. vi, p. 335.—Phil., Abbild., iii, t. 1, f. 2 (not *P. cymbularia* Delessert, Rec. de Coq. t. 23, f. 8, =*P. ænea* Martyn).—*P. cymbuloides* Lesson, Voy. de la Coquille, p. 422.—*Nacella cymbalaria* (sic) Rochebrune & Mabille, Mission Cape Horn p. 97.—*Nacella compressa* Mabille & Rochebrune, *l. c.* p. 98, t. 5, f. 9.—*P. cymbium* Phil., Arch. f. Naturg. 1845, p. 60; Abbild. p. 7.—*P. vitrea* Phil., Abbild., t. 1, f. 4.—*Nacella vitrea* Thiele, Das Gebiss, ii, t. 28, f. 28 (dentition).—*P. hyalina* Phil., *l. c.*, f. 3.—*Nacella hyalina* Thiele, *l. c.*, f. 29 (dentition).—*Nacella mytilina* Dall, Am. Journ. Conch. vi, p. 274, t. 16, f. 26 (dentition). Thiele, *l. c.*, t. 28, f. 30, (dentition).—*P. (Nacella) mytilina* Smith, Philos. Trans. clxviii, p. 181.

The thin texture, oblong form and anterior apex are diagnostic of this delicate species. It varies considerable in the position of the apex; in some specimens it is nearly marginal. The color is also variable, "some being of a general grayish tint, varied at intervals with darker concentric rings and often a few radiating palish stripes on the ribs. Others are uniformly yellowish-brown, others pale luteous broadly striped with black, and finally, others are of a uniform pale horny color, but all have the apex cupreous." (*S.*)

There is no reason for ignoring Helbling's work on this species except that his book is not a common one. But surely if Martyn's names are to be accepted, one cannot close the door upon the properly proposed names of the German author.

The synonymous *P. compressa* Mab. & Rochebr. is figured on pl. 50, fig. 37.

Var. HYALINA Phil. Pl. 50, figs. 38, 39.

Apex at or very near the anterior margin. The specimens before me show numerous forms between this and the typical *mytilina*.

A large series of radulæ must be examined before the differences found by Thiele can be acknowledged to be of specific value. This organ no doubt varies just as the shells do, in minor characters. Indeed there is often considerable variation in the teeth of a single odontophore!

Section *Patinella* Dall, 1871.

Patinella Dall, Amer. Journ. Conch. vi, p. 272.—Thiele, Das Gebiss der Schnecken, ii, p. 330, 1891.

This section differs from *Nacella* s. s. in having the branchial processes equally developed all around, and the shell more solid and more intensely colored.

The dentition is in all essentials similar to that of *Nacella* s. str. Indeed there are no characters upon which a separation more than specific can be based. At the same time, there is some considerable variation observed among the various species of *Patinella*, part of which may be specific, part merely individual variation. A large number of radulæ must be examined before safe specific characters can be based upon this organ, as it varies just as do the shells. Pl. 74, fig. 5, represents the dentition of *N. venosa ;* pl. 74, fig. 7, 8, that of *N. fuegiensis.*

The New Zealand species referred by Hutton to *Patinella* belong to *Helcioniscus (q. v.)*. They have the branchial cordon incomplete in front, and no epipodial ridge has been observed in them,—this last character being the main diagnostic mark of *Nacella + Patinella.*

N. ÆNEA Martyn. Pl. 15, figs. 5, 6 ; pl. 45, figs. 22, 23.

Shell solid, elevated, ovate, the breadth three-fourths of the length ; apex somewhat anterior, but behind the anterior third, and always somewhat curving forward. Sculptured by 34–38 rather strong radiating ribs, which are typically closely scale-grained, but as frequently almost smooth. Color brownish-ashen with several concentric dark brown zones; when worn, of a ferruginous-brown all over.

Interior lustrous, whitish or of a bronze tint, the center with an irregular red-bronze area, which is generally (but not always) rather small and indistinct. Margin scalloped.

Length 63, breadth 48, alt. 26 mill.

Straits of Magellan.

P. ænea MARTYN, Universal Conchologist i, t. 17.—REEVE, Conch. Icon., f. 9.—SMITH, Philos. Trans. vol. 168, p. 179, 1879.— ? *P. rustica* PERRY, Conchology, t. 43, f. 2.—*P. guadichaudi* BLAINV, Dict. Sci. Nat. xxxviii, p. 93 (1825.)

Typical *ænea* is thicker than var. *deaurata ;* the dark stripes of the exterior are faint or imperceptible within ; the apex is more central. It is more ovate than var. *magellanica*, the apex less erect.

The description and figures given above, as well as the synonymy, apply to typical *ænea* only.

The series of this species before me is very extensive, and shows such remarkable variations that it is not unlikely that all described species of *Patinella* will be found to be connected by intermediate forms. The specimens of *ænea* may be separated by moderately definite characters into three varietal types: (1) typical ÆNEA, defined above; (2) var. DEAURATA; and (3) var. MAGELLANICA.

Var. DEAURATA Gmelin. Pl. 46, figs. 28–36.

Shell rather thin, oblong, depressed, the apex in front of the anterior third, somewhat curving forward.

Interior showing bronze-brown radiating stripes on a light ground, the central area reddish or blackish bronze, oblong, generally distinct. Length 56, breadth 38, alt. 16 mill.

Straits of Magellan.

P. deaurata GMEL. Syst. Nat. xiii, p. 3703.—*P. cymbularia* DELESSERT (not Lam.) Rec. de Coq. Lam'k., t. 23, f. 8.—*P. ferruginea* SOWB., Genera of Shells, *Patella*, f. 4.—*P. delessertii* PHIL., Abbild. iii, p. 9, t, 1, f. 5.—*P. varicosa* REEVE, Conch. Icon. f. 21.— ? *P. adunca* PERRY, Conchology, t. 43, f. 5.—*Nacella strigatella* ROCHEBR. & MAB., Mission du Cap Horn, Moll., p. 96, t, 5, f. 8.— *P. ferruginea* WOOD, Index, *Patella*, f. 32.

Distinguished from typical *ænea* by being more depressed, thinner, the apex more anterior. The riblets of the outer surface are sometimes nearly obsolete, sometimes strongly developed; and they may be either smooth or granose, the grains having a scale-like character, as in *ænea*.

I am much disposed to consider *P. polaris* of Martens and Pfeffer a form of this variety.

Specimens in which the stripes anastomose and branch were called *varicosa* by Reeve (pl. 46, figs. 33). I surmise that Gmelin's *P. flaminea* (Syst. xiii, p. 3716) and Woods' *P. flaminea* (Index, pl. 38, f. 71) are identical with *varicosa*.

The outer surface is frequently bluish-white, ribs yellowish, but sometimes the ribs are rust-brown. I have seen specimens of a clear yellow, lacking radiating stripes; and others occur in which the stripes unite to make a uniform dark brown shell.

This variety is further modified into oblong forms having the apex decidedly curved over, and near to the anterior end. Such a form is that called *cymbularia* by Delessert (pl. 44, fig. 20), and *striga-*

tella Rochebrune and Mabille. These forms are like *Nacella mytilina* in contour, but they are more solid and ribbed.

Var. MAGELLANICA Gmelin. Pl. 44, figs. 9–17; pl. 43, figs. 1–6.

Shell rounded-oval, high-conical, the apex nearly erect; varying from strongly radiately ribbed to smooth. Unicolored, radiately streaked, or having oblique stripes.

Interior generally very dark, the muscle-scar sometimes snowy-white. Length 45, breadth 36, alt. 27 mill.

P. magellanica GMEL., Syst. xiii, p. 3703.—REEVE, Conch. Icon., f. 19.—*Patinella magellanica* DALL, Amer. Journ. Conch. vi, p. 273, t. 15, f. 24 (soft parts and dentition).—*Patella atramentosa* RVE., Conch. Icon., f. 41.—*P. venosa* Rv., f. 18.—*P. chilænsis* Rv., f. 98.—*P. meridionalis* ROCHEBRUNE & MABILLE, Bull. Soc. Phil. Paris 7th Ser., ix, p. 109, 1885; Mission Scientifique du Cap Horn, vi, p. 94, t. v, f. 4.—*P. metallica* R. & M., *l. c.*, p. 91, t. 5, f. 5.—*P. pupillata* R. & M., *l. c.*, p. 92, t. 5, f. 6.—*P. tincta* R. & M., *l. c.*, p. 93, t. 5, f. 7.

Rounder than typical *ænea*, and having a more central, erect and elevated apex.

Here belong a number of forms described by Reeve years ago, and by Rochebrune and Mabille recently. The paper by the last authors, on the mollusks of Cape Horn, is an admirable specimen of how systematic zoological work should *not* be done. The ignorance displayed is only excelled by the lack of judgment. We should, however, give MM. Rochebrune and Mabille the benefit of the doubt as to whether their species and groups were intended seriously or as a jest.

. This variety is well represented by the figures 9–11 of pl. 44, drawn from specimens collected at Santa Cruz River, Patagonia. The ribs are strong and carinated in some specimens, almost completely obsolete in others, this comparative smoothness not being the result of erosion. The central area of the interior is sometimes partly of a snow-white color (pl. 44, fig. 12); sometimes the muscle-scar is white (pl. 44, fig. 16.)

P. meridionalis R. & M. (pl. 43, figs. 1, 2), *P. pupillata* R. & M. (pl. 43, figs. 3, 4), *P. tincta* R. & M. (pl. 43, figs. 5, 6) and *P. metallica* R. & M. (pl. 44, figs. 17) are ordinary forms of *magellanica.*

The form called *atramentosa* by Reeve (pl. 44, figs. 13, 14,) has the ribs wide and subobsolete; whitish with broad blackish rays.

Reeve's *P. chiloensis* (pl. 45, figs. 20, 21,) is similar to *atramentosa* but darker.

The *P. venosa* Rve. (pl. 45, figs. 24–27) is rounded, the ribs almost completely obsolete, painted with divaricating stripes. The interior is peculiarly rich in coloring, being suffused with bronze-red, the muscle-scar lilac. Numbers of this form are before me.

I suppose the *P. areolata* Gmel. (Syst., p. 3716; Wood's Index, pl. 38, f. 70) to be the same as *venosa*.

N. INRADIATA Reeve.   Pl. 20, figs. 43, 44.

Shell ovate, elevately convex, somewhat compressed at the sides, everywhere rather obsoletely, radiately latticed; whitish, obscurely rayed with a very few red lines in pairs, lines irregular, distorted. Interior white, red lines distinct, chestnut in the middle.   (*Rve.*)

Habitat unknown.

P. inradiata REEVE, Conch. Icon., f. 77.   Jan., 1855.

A deep cup-like opaque-white shell, obscurely latticed throughout, irregularly rayed with a few straggling red lines, which are most distinctly seen in the interior.   (*Rve.*)

This is a form I have not seen. It may prove to be one of the many color-varieties of *P. ænea var. deaurata.*

N. POLARIS Hombron & Jacquinot.   Pl. 49, figs. 21–27.

Shell oblong-ovate, more or less compressed, sculptured with rounded, rather distant, sparingly nodose radiating ribs, often obsolete; margin roundly crenated; exterior usually eroded brown; interior shining bronze-black; vertex inclined forward, situated at the front $\frac{1}{4}$–$\frac{2}{3}$ of the length.   (*M.*)

Length 51, breadth 36, alt. 23 mill.

Length 48, breadth 35, alt. 26 mill. (especially high.)

Length 59, breadth 40, alt. 22 mill (the longest, rather flat.)

Length 47, breadth 33, alt. 14 mill. (the flattest.)

South Georgia.

P. polaris MARTENS & PFEFFER, Mollusken von Süd Georgien, in Jahrb. der Hamburgischen Wissenschaftlichen Anstalten iii, p. 101, t. 2, f. 11–13 (1886).—*P. polaris* H. & J., Ann. des Sci. Nat. (ii) xvi, p. 191, (1841.)

P. kerguelensis E. A. Smith is notably wider posteriorly, and is not so dark inside.   (*Mart.*)

P. polaris approaches certain forms of *P. œnea* var. *deaurata*, but the ribs are more obsolete and the interior darker than is usually the case in that form.

N. KERGUELENSIS E. A. Smith. Pl. 43, figs. 7, 8.

Shell oval, a little narrowed in front, rather elevated convex, the apex prominent and well toward the front, especially in the young; widely radiately ribbed, the ribs little prominent, and often with interstitial smaller riblets; sculptured with elegantly undulating and close concentric growth-lines. Exterior bluish-ashen, the ribs usually darker, toward the apex ferrugineous in eroded examples. Interior bronze-brown, generally paler toward the margin, which is a little undulating; muscle-scar visible; a large example measures, Length 82, breadth 70, alt. 45 mill. (*Smith*.)

Swain's Bay and Royal Sound, Kerguelen Island, in about 1 fm.

P. ferruginea (SOWB. MS. in Mus. Cuming) REEVE, Conch. Icon., f. 40, not *P. ferruginea* Gmel. nor *P. ferruginea* Sowb. Genera of Shells, f. 4.—*P. kerguelensis* E. A. SMITH, Zool. of Kerguelen Id., Moll., in Philos. Trans. Roy. Soc. Lond. vol. 168, p. 177, t. ix, f. 13, 13a, (1879.)

Differs from *P. œnea* Martyn in having the shell less prominently costated, differently colored, and in the apex being very prominent and much curved over, so as to give it a capuliform appearance, a character constant in all specimens, young and old, elevated and depressed; it also differs in the coloration of the animal. (*Smith*.)

N. FUEGIENSIS Reeve. Pl. 49, figs. 28, 29, 30, 31.

Shell oval, rather thin, semitransparent, rather compressly raised; apex rounded, inclined anteriorly; radiately ridged, ridges thin, numerous, densely crossed with concentric striæ; greenish, more or less stained and blotched with chestnut-brown, apex bronze; interior iridescent-bronze, radiately grooved, grooves sometimes partially obsolete. (*Reeve*.)

Tierra del Fuego; Orange Bay; Falkland Is.; Royal Sound and Swain's Bay, Kerguelen Id., everywhere common on the submerged fronds of long floating kelp (Macrocystis) bordering the shore.

P. fuegiensis REEVE, Conch. Icon., f. 73, (1855).—*Patella (Patinella) fuegiensis* SMITH, Philos. Trans. vol. 168, p. 180, t. 9, f. 14, 14a.—*P. fuegiensis* ROCHEBRUNE & MABILLE, Miss. Sci. du Cap Horn, Moll., p. 95.

"The description given by Reeve is very good, but he does not lay sufficient stress upon the beautiful raised concentric ridges. He calls them striæ, which term scarcely gives the idea of thread-like lirations such as these. They are very closely packed and undulate very prettily on and between the numerous radiating ribs.

"The figure, except in outline and the position of the apex, gives but a poor idea of this beautifully sculptured *Patella*. It represents the number of ribs at about forty, whereas there are usually about sixty. The specimens from Kerguelen's Island are a trifle narrower and much more depressed than examples from the Falkland Islands; in fact, it is only near the apex that they are at all raised, and toward the margin they are upturned, so that the dorsal surface is concave, and this form of the shell certainly prevents the animal from entirely concealing itself when adhering to a flat surface; But this peculiarity of form only exists in adult specimens, for several small ones are like ordinary species in this respect. The radiating ribs are almost obsolete in the flat examples, but the undulating concentric lirations, which are more prominent and farther apart than in the type form of the species, define their position; in young shells they are more pronounced. Color generally uniformly purplish slate, with the apical region ferrugineous; interior similarly tinted, but rather more deeply. One shell has a white border. They are all very thin and fragile, and the edge is very liable to break off in a line with the concentric raised lines of growth.

"The animal has the sides and sole of the foot greenish-grey, the edge of the mantle and gills pale buff, the tentacular filaments on the margin of the mantle blackish except at their tips, tentacles short and thick, pale buff, with a black spot above.

"The frill-like expansion of the foot, similar to that of *P. ænea* and *P. kerguelensis*, is a little above its edge, is bluntly serrated, and interrupted beneath the head.

"Teeth of the lingual ribbon slightly hooked, in pairs, scarcely diverging; the central pair two-pronged, the inner prong much larger, spear-head shaped; the lateral pairs alternating with the central ones are four-pronged, the innermost prong smallest, the next two subequal, and the outside one situated nearly at right angles to the rest of the tooth, about the same size or a trifle larger." (*Smith.*)

N. CLYPEATER Lesson. Pl. 50, figs. 40, 41, 42, 43.

Shell circular or rounded-oval, rather thin but solid, depressed, the apex a little in front of the middle. Outer surface lusterless, sculptured with fine and rather close radiating riblets, sometimes subobsolete. Brownish or tawny, the riblets often lighter, usually mottled with whitish toward the apex.

, Interior bright silvery, the central area of a deep red-bronze, muscle-scar snowy-white, surrounded with red-bronze.

Length 61, breadth 58, alt. 14½ mill.

Coast of Chili at Valparaiso and Saint Vincent.

P. clypeater LESSON, Voy. de la Coquille, Zool., p. 419, (1830).—REEVE, Conch. Icon., f. 38, 38b.

A nearly circular, depressed shell, having the bronze-brown coloring characteristic of *Patinella*. The interior sometimes has no white horse-shoe; and in some examples the silvery outer zone is considerably invaded by bronze stains.

This species has been reported from California, Lower California and Japan, but not correctly.

Genus HELCIONISCUS Dall, 1871.

Helcioniscus DALL, Amer. Journ. Conch. vi, p. 227, (type *Patella variegata* Reeve).—THIELE, in contin. Troschel's Das Gebiss der Schnecken, ii, pt. 7, 333 (full discussion and figures of the dentition). —*Cellana* H. ADAMS, P. Z. S. 1869, p. 274, (type *N. cernica.*)

The gill-cordon is interrupted in front.

There are no epipodial processes or ridge on the sides of the foot.

The formula of teeth is $_3(_1^{111}_1)_3$.

The radula is long and spirally rolled. The rhachidian tooth is narrow, with a variously shaped forward appendage. The inner lateral has an outward wing, and a simple, long, cusp; the outer lateral has an inner long point with an outer side-cusp, besides a short, mostly rounded longitudinal cutting-edge. The marginal teeth are characteristic in appearance, the posterior part being divided from the anterior, the connection being so thin as to be imperceptible. The inner marginal tooth has a cusp on its front end (pl. 74, fig. 6, *P. capensis* Gmel.)

The shell is conical, apex subcentral or subanterior; inside having a silvery and mica-like luster.

Distribution, Indian and Pacific Oceans, but not found on the American shores north of Chili. No species have been found in the Atlantic Ocean.

This group is closely allied to the *Patinella* section of *Nacella*, having a very similar radula. It differs in the lack of an epipodial ridge, in having the gill-cordon interrupted, and in the silvery-

micaceous rather than bronzed-*metallic* luster of the interior of the shell.

I have adopted below a division of the numerous species of *Helcioniscus* into geographic groups as follows:

1. Chilian species.
2. Polynesian and East Indian species.
3. Japanese and Chinese species.
4. Species of New Zealand and Australia.
5. Species of East Africa from the Red Sea to the Cape, and the adjacent islands.
6. (Species of unknown locality.)

Without having examined all of the species, a perfectly natural zoological grouping would be impossible. Certain apparently natural groups, however, force themselves upon us. Such are the *group of P. sagittata*, containing *sagittata, tahitensis, ardosiœa, amussilata, toreuma, nigrolineata.*

The *group of P. exarata*, containing *exarata, argentata, melanostoma, boninensis, nigrisquamata, stearnsii, grata.*

The *group of P. rota*, containing *rota, capensis, novemradiata, variabilis, dunkeri, cernica, eucosmia, garconi*, etc.

The New Zealand and Australian species also form a natural group of equal value with the preceding.

A number of outlying forms, as well as some described species the shells of which I have not seen, apparently belong to neither of these four assemblages.

(1). CHILIAN SPECIES.

The limpets of South America belong mainly to *Acmæidæ.* The *Patellidæ* being represented by *Nacella* + *Patinella* and by one or two forms of *Helcioniscus.* Of these, the habitat of *P. ardosiœa* is unquestionable, but that of *P. nigrisquamata* may still be considered open to revision.

H. ARDOSLEUS Hombron & Jacquinot. Pl. 32, figs. 63, 64, 65, 66.

Shell erectly conical, short-oval or nearly circular, the apex nearly central; *slopes straight.* Sculptured with close radiating striæ, which are *not granose*, every fourth one being somewhat larger. *Slate colored or light bluish-olive*, darker on the upper part of the cone, the eroded apex usually reddish.

Interior bluish-white, with silvery and opalescent reflections, the central callus opaque-white tinted more or less with reddish, the

front segment often darker, the edges more or less leaden.   Edge
of shell narrowly bordered with slate-color.

Length 58, breadth 52, alt. 19 mill., (fig. 65.)

Length 48, breadth 40, alt. 24 mill., (fig. 66.)

Island of Juan Fernandez.

P. ardosiæa H. & J., Ann. des Sci. Nat. (2), xvi, p. 190, (1841).
—*P. clathratula* REEVE, Conch. Icon., f. 30, (1854.)

This is a very distinct form, allied to *P. tahitensis* Pse. but not
granulose (although there are concentric striæ), and not variegated.
There are often very fine concentric wrinkles inside ; and the interior
sometimes has an opaque white or lemon-tinted peripheral zone, or
rarely the entire surface between the reddish spatula and the narrow
border is silvery-yellow.

Half-grown specimens greatly resemble Hutton's *P. olivacea*
except that that species has the apex slightly more anterior, the
spatula broader in front and not reddish, and the border narrower
and blacker.

This is one of the few species of Hombron and Jacquinot's paper
which is portrayed by them with unmistakable accuracy.   The
specimens orginally described and those of Reeve, as well as the
originals of my figures 64, 65, 66, were from Juan Fernandez.

H. NIGRISQUAMATUS Reeve.   Pl. 19, figs. 35, 36 ; pl. 48, figs. 13, 14,
15.

Shell solid, oval, erectly, *straightly conical*, the altitude about half
the breadth ; apex erect, acute, a little in front of the center.   Sculptured with *strong, narrow, closely nodule-scaled ribs*, obscurely or
obviously alternating in size, and about 46–51 in number.   Color
yellowish, often blackish-gray above and orange around the circumference, *having here and there scattered black scales or nodules.*

Interior silvery, more or less orange-tinted and decidedly iridescent, the central area small, dark chestnut colored.   In young or
half-grown shells *the interior shows black spots corresponding to the
black nodules of the outer surface*, but in adults they are obscured by
the enlarging central callus.   Edge having a narrow distinct fleshy-
orange border.

Length 63, breadth 52, alt. 28 mill.

Length 83, breadth 71 mill.

Concepcion, Chili (Dr. Ruschenberger) ; *Australia* (Reeve.)

Patella nigrisquamata REEVE, Conch. Icon., f. 3, (1854).—
Patella mazatlandica SOWB., Zool. Beechey's Voy. Moll., p. 148, t.
39, f. 12, 1831.

This species may be known by its erectly conical form, narrow
ribs, having close, high solid scales or nodules, of which some are
black, especially upon the upper part. The central area is small,
brown. The young (pl. 54, fig. 13) are spotted inside.

The *P. nigrisquamata* of Reeve, is a shell exhibiting but little
variation. Reeve gives the locality "Australia," but this has not
been confirmed. One tray of the specimens before me are from
Concepcion, Chili, collected by Dr. W. S. W. Ruschenberger. The
others lack record of locality.

The *P. mazatlandica* of Beechey's Voyage, was said to come from
Mazatlan, but this is of course a mistake. Carpenter at one time
referred it to one of the Californian *Acmæas* as a synonym, but later
concurred in Hanley's opinion that it was the same as *P. exarata*
Nutt. This opinion I am not prepared to endorse, for the internal
central callus of *P. exarata* is white, or sometimes partly black or
violet-purple, but it is never, I believe, *chestnut brown*, as in the
figures and description of *Mazatlandica*, which correspond exactly
with young specimens of *nigrisquamata*, now before me.

The name *mazatlandica* has priority, but in view of the doubt
attaching to its use, and the fact that it is a misnomer, I have deemed
it wisest to retain Reeve's designation.

Compare also *Patella boninensis* Pilsbry.

The scattered black spots found on the upper part of the cone are
also seen in some specimens of *P. exarata*, and some other allied
species.

(2). EAST INDIAN AND POLYNESIAN SPECIES.

This region possesses species of two groups: strong, conical, ribbed species, such as *exarata*, *argentata;* and lower, more delicately
sculptured forms, *sagittata*, *tahitensis*, *testudinaria*. The Chilian
forms are evidently derived from this fauna, and the Japanese
species are very closely allied, belonging to the same two groups.

H. EXARATUS Nuttall. Pl. 47, figs. 1, 2, 3, 6, 7, 8, 9, 10, 11, 12.

Shell oval, conical, the apex slightly in front of the center; slopes
nearly straight. Surface sculptured with many very strong black
ribs on a slightly lighter ground, the ribs subequal, about 38–48 in

number; a few small intervening riblets often developed. Color typically almost black.

Interior somewhat silvery, leaden-bluish, showing the dark ribs; central area black, whitish in front, sometimes entirely white; often of a beautiful purple.

Length 40, breadth 34, alt. 16 mill.

Sandwich Islands.

Patella exarata NUTT., Jay's Catal., p. 38; REEVE, Conch. Icon., f. 47a, 47b, and 62a, 62b, 1854.—PEASE, Amer. Journ. Conch. vii, p. 198, 1872.—*P. sandwichensis* PEASE, P. Z. S. 1860, p. 537; Amer. Journ. Conch. vii, p. 198, 1872.—*P. undato-lirata* REEVE, Conch. Icon., f. 59.—*Helcioniscus exaratus* DALL., Amer. Journ. Conch. vi, p. 279, t. 16, f. 29 (dentition).—*Patella lugubris* Blainv., DESH. Trait. Elém. de Conchyl., Atlas, t. 62, f. 13, 14. (*non* Blain.)

This common *Patella* of the Sandwich Is. is readily known by its numerous strong black ribs. It is so variable, and intermediate forms are so numerous, that varietal names are scarcely admissible.

Some individuals (figs. 6, 7) are drab or gray, closely spotted with black on the upper part of the cone, and having the ribs nodulose, as in *Patella nigrisquamata.*

This species has been reported by Schrenck from the bay of Hakodadi, Japan. His specimens were identified by a comparison with Reeve's figures, not by comparison with actual specimens. In all probability the Japanese specimens were *imported*, as were those said by Reeve to be from Oregon.

The typical *exarata*, as first defined and figured by Reeve, is dark, with black ribs (figs. 8, 9, 10.)

Pease states that his *sandwichensis* differs from typical *exarata* in being thicker, more elevated, ribs generally larger, wider apart, more prominent, and crenate or scabrous, generally having smaller intermediate riblets. It is generally white inside. It inhabits deeper water, and according to Mr. Pease, differs in *taste* from typical *exarata!*

The form called *lutrata* Nutt. (pl. 47, figs. 1, 2, 3) is white, with a few of the ribs brown. The central callus of the interior is white.

H. ARGENTATUS Sowerby. Pl. 18, figs. 29, 30 ; pl. 65, fig. 93.

Shell large, solid, of a *dome-like conical* form, oval; slopes convex ; apex subcentral (in the young more or less anterior). Sculptured with very numerous unequal radiating riblets. Color chestnut-brown, becoming coppery when rubbed; apex usually eroded and coppery.

Inside having a large, distinct central callus of pure white, a lusterless whitish zone around the muscle-impression, outside of which it is bright, shining, and silvery or of a pale golden tint.

Dimensions of a moderate-sized individual: length 90, breadth 79, alt. 42 mill.

Sandwich Is.

Patella argentata SOWERBY, in Zool. Beechey's Voy., H. M. S. Blossom, Moll., p. 148, t. 39, f. 7, (1839).—? *Helcioniscus ? argentatus* "Gray" DALL, Amer. Journ. Conch. vi, p. 278, (This may be *P. clypeater* Less.).—*Patella cuprea* REEVE, Conch. Icon., f. 15, Oct., 1854, (See also, errata to *Patella*, at end of index, Conch. Icon.).— *Patella talcosa* GLD., U. S. Expl. Exped. Moll. & Sh., p. 334, atlas f. 452.—*P. opea* NUTTALL, according to specimens deposited in Mus. Acad. N. S. Phila. by Nuttall (not *P. opea* Rve.)

A large, dome-shaped species, brown outside, pale-golden with a white central callus inside. This species has been reported from Australia, Chili and Japan, but upon wholly insufficient evidence.

In Beechey's Voyage, the species is said to be from Chili, but as the expedition also touched at Tahiti, and many of their shells became mixed, there is no doubt of the true locality whence the types were brought. The description and figure are unmistakable. Several writers on the Japanese fauna have confused this with *P. clypeater*, but this species is more raised than that, more solid, and *never* marked with brown inside.

H. ARTICULATUS Reeve. Pl. 65, figs. 87, 88.

Shell hexagonally ovate, attenuated in front, rather flatly depressed, everywhere radiately ridged and striated; olive, yellow rays at the angles, ridges articulated with purple-black and white. Interior rather silvery.

A rather compressly flattened species, divided on the surface into seven to nine subangular areas, rayed with neatly articulated ridges. (*Rve.*)

Island of Ticao, Philippines.

P. articulata REEVE, Conch. Icon., f. 97, 1855.

H. TESTUDINARIA Linné. Pl. 25, figs. 16, 17, 18, 19.

Shell large, *thick and solid*, oval, conic or depressed, the apex at about the front third; posterior slope convex. *Surface nearly smooth*, but showing inconspicuous, close, low radiating riblets, generally more obvious in the young. Broadly rayed with alternate dark chestnut and soiled white or pinkish, the dark rays spotted with light, the light rays with dark.

Interior varying from bluish-white to yellow, somewhat translucent and with a micaceous luster, the central area white, often outlined with brown or yellow; border colored by the dark rays of the exterior.

Length 85, breadth 75, alt. 33 mill.

Length 87, breadth 74, alt. 24 mill.

Philippines (Rve.); *Singapore* (Phil. Acad. Coll.) ; *Cape Good Hope* (Frauenfeld.)

P. testudinaria LINN., Syst. Nat. x, p. 783.—HANLEY, Sh. of Linn., p. 427.—REEVE, Conch. Icon., f. 6.—*P. insignis* DUNKER, Verh. Zool.-bot. Gesell. Wien, 1866, p. 941.—FRAUENFELD, Reise der österreichischen Fregatte 'Novara' um die Erde, Zool. Theil, ii, Moll., p. 14, t. 2, f. 25, 1868.—*P. rumphii* BLAINV. Dict. Sci. Nat. xxxviii, p. 95, (1825.)

A large solid species, nearly smooth, rayed and curiously spotted with brown and whitish.

H. TAHITENSIS Pease. Pl. 67, figs. 4, 5, 6, 7, 8.

Shell conical, rounded-oval, *the apex erect,* and situated *within the middle third* of the length. Slopes but slightly convex. Surface sculptured with closely granulose lirulæ, indistinctly alternating in size. Color dusky, with indistinct reddish zigzags ; but if held toward the light a beautiful pattern of *distinct reddish zigzag stripes* on a light ground, is seen.

Interior silvery-blue, *the central callus white in front,* becoming indistinct and *leaden* posteriorly.

Length 34, breadth 30, alt. 11 mill.

Tahiti.

Tectura tahitensis PSE., Amer. Journ. Conch. iv, p. 98, t. 11, f. 21, (1868.)

Decidedly more erect than *P. sagittata,* with coarser radiating lirulæ and different coloring. It is more coarsely sculptured than *P. amussitata.* Fig. 8 represents the color-pattern as seen by looking through the shell.

The zigzag or v-shaped markings are much broken in one specimen before me (fig. 7), and there are several broad red dashes on one side. A large series would doubtless show further variations and interruptions of the typical pattern. 9

H. SAGITTATUS Gould.   Pl. 65, figs. 89, 90, 91, 92.

Shell depressed, rounded-oval, *the apex curving forward*, situated at about the *front fourth;* surface closely sculptured with radiating striæ, which are granose in front, usually less so behind. "Color dusky olive-green, with obscure radiations of dusky; but if held up to the light the whole shell is found to be most beautifully reticulated and radiated with arrow-head dusky spots, often articulated with whitish."

Interior blue or bluish-white, the central callus orange in adults, yellowish-green in the young, its outlines not sharply defined.

Length 38, breadth 34, alt. 13 mill.

Length 39, breadth 36, alt. 9 mill.

Viti and Fiji Islands.

P. sagittata GLD., Proc. Bost. Soc. N. H. ii, p. 148, July, 1846; U. S. Expl. Exped. Moll. & Sh., p. 337, atlas, f. 449.

A rounded-oval, depressed and arched species, having finely granose radiating lirulæ. The coloring is dusky, indistinctly patterned, but if held toward the light an elegantly figured design is seen. There are usually separated narrow radii of alternate dark and light, the intervals variously reticulated; but these articulated rays are often lost in the general dusky net-work.

This species differs from *P. amussitata* in having the apex more anterior, the ribs finer, the pattern of coloring different. It is not so elongated as *P. toreuma* and differs in color-pattern and sculpture.

H. REYNARDI Deshayes.   Pl. 66, figs. 94, 95.

Shell ovate, thin, fragile, depressed, radiately striated, striæ unequal, subgranose; apex obtuse; margin entire; interior pearly, central disc white; outside having radiating reddish-brown bands.

Length 50, width 40 mill.   (*Dh.*)

Ceylon.

P. Reynaudi DH., in Belangers' Voy. aux Indes-Orientales, Zool., p. 411, atlas t. 2, f. 11, 12; DH. in Lam'k., vii, p. 543.

H. FLEXUOSUS Quoy & Gaimard.   Pl. 66, figs. 96–98.

Shell small, fragile, orbicular, angulate, arcuate, the vertex only elevated; margins flexuous, obsoletely striated, whitish-brown dotted, apex rosy; inside bluish, cavity orange.

Length 10 lines.   (*Q. & G.*)

Island of Vanikoro.

P. flexuosa Q. & G., Voy. de l'Astrol. iii, p. 344, pl. 70, f. 9–11.

(3). JAPANESE AND CHINESE SPECIES.

The Patellas of this region may be distinguished by the key given below.

Neither *P. clypeater* nor *P. argentata* (*cuprea*) inhabit Japan, the citations of these species by Schrenck and Dunker being erroneous.

Key to Japanese Patellidæ.

Shell more or less silvery or iridescent inside (*Helcioniscus.*)
    Conical, having about 50 strong, close elevated riblets, alternating
            [or subequal in size.
        Large, light buff; border of the inside narrow, yellowish,
                H. Boninensis Pils.
        Variegated with brown; border of the inside conspicuously
            black-blotched,   *H. Stearnsii* Pils.
    Conical, having several smaller riblets in each interval between
        the larger ribs,   *H. pallidus* Gld.
    Ribs fine or obsolete.
        Solid, with radiating dark lines; spatula orange-red; ribs
            obsolete,   *H. nigrolineatus* Rve.
        Thin, with very finely beaded riblets or striæ,
                H. amussitatus Rve.
        Thin, with very fine striæ, not beaded,   *H. toreuma* Rve.
Shell porcellanous inside, opaque, not iridescent (*Scutellastra.*)
    Depressed, having strong irregular ribs,   *P. stellæformis* Rve.

H. BONINENSIS Pilsbry. Pl. 66, figs. 1, 2; pl. 67, fig. 3.

Shell large, solid, oval, erectly, straightly conical, the apex a little in front of the middle; posterior slope a little convex, the other slopes straight; basal side-margins a little elevated, so that the shell is supported by the ends alone when resting upon a plane surface.

Sculptured with numerous (48–55) subequal radiating ribs (and sometimes some small interstitial riblets), the ribs varying from closely and rather weakly crenulated to strongly tubercled.

The color is a uniform light buff, somewhat soiled, and having sometimes a few small black spots around the apex.

Interior: the muscle-impression is of a snowy or bluish-white; outside of it there is a broad band varying in different individuals from creamy-brown to deep chestnut, outside of which there is a silvery, slightly iridescent zone, extending to the narrow yellow border. The large central area is either cream-colored with a distinct dark laciniate outline, or is of an umber-brown, lighter in the depth. *From each angle of the anterior head-segment of the central callus, a narrow dark band radiates,* passing through the dark zone which surrounds the muscle-scar.

Length 100, breadth 85, alt. 40 mill. (no. of riblets 53.)
Length 90, breadth 75, alt. 40 mill. (no. of riblets 50.)
Length 88, breadth 77, alt. 40 mill. (no. of riblets 48.)

Bonin Is., north of Japan.

Patella (Helcioniscus) boninensis PILSBRY, The Nautilus, November, 1891, p. 79.

This magnificent limpet approaches the *P. nigrisquamata* of Reeve, but may be readily distinguished by its much larger central callus inside, the two diverging brown streaks mentioned in the description, etc. The two forms are moreover widely separated geographically.

The specimens were seen and purchased by Mr. Frederick Stearns, of Detroit, Michigan, at the Third National Exhibition at Tokyo, 1890. They are called in Japanese, *Yome-gai-sara,* "Bridecup shells."

H. STEARNSII Pilsbry. Pl. 48, figs. 16, 17, 18.

Shell solid, elevated-conical, oval; apex a little behind the front third of the length; front slope straight or concave, posterior slope convex. Sculptured with about 51 unequal closely nodulose ribs, separated by deep interstices. Surface lusterless, soiled whitish, having irregular rays of reddish-brown, and speckled with the same on the upper part of the cone.

Interior bluish-white rayed or mottled with darker by the dark rays of the exterior; central area *strongly defined, reddish-brown* with a white stain in the cavity of the apex; edge of the shell scalloped. The dark rays become vivid deep brown or black at the border, the dark blotches alternating with white.

Length 41, breadth 31½, alt. 27 mill.
Length 38, breadth 29, alt. 21 mill.

Province of Kii, Japan.

P. (Helcioniscus) Stearnsii PILSBRY, The Nautilus, Jan., 1891, p. 100.

This handsome shell is sculptured with closely nodulose ribs, obscurely alternating in size. There are 10 or 11 irregular dark rays on the outside, much broken into spots on the upper part of the cone. The apex is at the front ⅓ of the length. The interior shows fine transverse wrinkles when examined under a lens. It is not especially iridescent except at the border, where there is a narrow strip of fiery-orange iridescence between the blue-white of the interior and the blackish stained border. The species is named in honor of Mr. Frederick Stearns of Detroit, Michigan, who collected the shells when traveling in Japan.

H. PALLIDUS Gould. Pl. 67, figs. 9, 10.

Shell subovate, elevated, erectly-conical, solid; white both outside and within, or whitish-yellow; having close concentric sulci and striæ, and radiating unequal ribs, the ribs plicate or obsoletely plicate-tuberculate; there are 20–25 larger ribs radiating from the summit itself; of the smaller interstitial riblets there are about 60. Apex subcentral or at the front ⅓ of the length. Margin of the aperture undulating. (*Schrenck.*)

Length 40, breadth 33, alt. 24 mill.

Length 60, breadth 51, alt. 29 mill.

Hakodadi, Japan.

P. *pallida* GOULD, Proc. Bost. Soc. N. H. vii, p. 162.—DKR., Ind. Moll. Mar. Jap., p. 156.—*P. lamanonii* SCHRENCK, Reisen und Forsch. im Amurl. ii, p. 303, t. 14, f. 6–9.

Characterized by its light color and the sculpture, consisting of radiating ribs, and having several riblets in each interval. I have seen none but immature specimens.

H. NIGROLINEATUS Reeve. Pl. 14, figs. 71, 72, 73, 74; pl. 13, figs. 48, 49.

Shell solid, oval, rather depressed; apex at the front third; slopes slightly convex; surface nearly smooth, the young having sub-obsolete radiating riblets; bluish, with numerous narrow radiating stripes of red, or sometimes black.

Interior dark-silvery, showing (especially in young shells) black radiating stripes. *Central callus bright orange-red, veined with black, its front portion white or nearly so.*

Length 74, breadth 60, alt. 18 mill.

Enoshima (Fr. Stearns), *Nagasaki* (Lischke) *and Tsus-sima* (Ad.), *Japan; Camiguan, Philippines* (Rve.)

P. *nigrolineata* REEVE, Conch. Icon., f. 43.—LISCHKE, Jap. Meeres-Conchyl. i, p. 111, t. 8, f. 5–11; ii, p. 103, t. 7, f. 1–6.

A magnificent species, the handsomest of the Japanese limpets. The rays of the outside are generally dull red, but sometimes are brown or dull black. Between them the surface is normally of a peculiar light blue tint. In some specimens (pl. 13, figs. 48, 49), there are numerous fine, close waved concentric reddish lines, and these shells generally show a peculiar mottled pattern inside. The spatula is normally orange-red, more or less veined with black, but the black sometimes predominates. There is, in all the specimens I have seen, a light or white tract on the forward part of the spatula.

My description is drawn from specimens kindly furnished me by Mr. Frederick Stearns.

Var. DIVERGENS Pilsbry.   Pl. 73, figs. 81, 82, 83, 84.

A *Helcioniscus* between *nigrolineata* and *toreuma*. It is rather thick and solid, at least as thick as *nigrolineata*. Surface having a distinct sculpture of subgranose, crowded unequal radiating striæ. Apex at the front fourth. Color purplish-brown, having rays of greenish-white, of which nine extend to the margin.

Interior dark, leaden, with silvery reflections, showing the white rays. Spatula opaque white in front, leaden-brown behind.

Length 42½, breadth 32, alt. 8½ mill.

Enoshima, Japan.

Fig. 84 represents the shell as it appears when held between the eye and a strong light. The color-pattern, otherwise obscure, is thus distinctly seen.

H. AMUSSITATUS Reeve.   Pl. 14, figs. 75–79; pl. 68, figs. 11, 12, 13.

Shell thin but rather solid, ovate, conical; apex in front of the middle; slopes nearly straight. Surface sculptured with fine, close, *regularly and closely beaded* radiating riblets or striæ. Light yellowish-brown, with inconspicuous darker rays, usually 11 in number, and sometimes speckled with reddish and opaque white, or mottled with purplish.

Interior bluish-silvery, iridescent, conspicuously finely crenulated toward the border; central area not distinctly outlined, whitish or dull brown. Length 44, breadth 36, alt. 12½ mill.

Bonin Is. and Japan to Philippines.

P. amussitata Rve., Conch. Icon. f. 83.—SCHRENCK, Reis. und Forsch. im Amurl. ii, p. 30, t. 14, f. 4, 5.—LISCHKE, Jap. Meeres-Conch. i, p. 109; ii, p. 100, t. 6, f. 7–11.—DKR., Ind. Moll. Mar.

Jap. p 156.—DEBEAUX, in Journ. de Conchyl. 1863, p. 245.—*P. granostriata* REEVE, Conch. Icon. f. 126.

This species is extremely variable in degree of elevation. It is never so variegated as *P. toreuma*, is more elevated, thicker, and the fine riblets are *distinctly, finely and regularly beaded*. From the Japanese species of *Acmæa* which have similar sculpture, it is of course separated by the lack of a defined internal border to the lip-edge.

I am now disposed to consider *Patella granostriata* Reeve as a synonym of *amussitata*.

H. TOREUMA Reeve. Pl. 13, figs. 50, 51, 52, 53.

Shell *depressed*, long-oval, thin, the apex between the front third and fourth of the shell's length; front slope straight or concave, posterior slope gently convex. Surface having fine close radiating striæ separated by interstices slightly wider than themselves, the striæ sometimes a little irregular but *not distinctly beaded*. Color excessively variable, usually greenish or buff, rayed or blotched with purplish-black and dotted with white; sometimes without dark markings.

Interior silvery or bluish, showing distinctly the dark and light markings of the exterior, the central area dusky, white or rich chestnut, *its edges not sharply defined*.

Length 40, breadth 31, alt. 9 mill.

Nagasaki, Tokio, etc., Japan; China.

P. toreuma RVE., Conch. Icon. f. 69.—LISCHKE, Jap. Meercs-Conchyl. i, p. 109, t. 8, f. 12–15 ; ii, p. 102, t. 6, f. 12.—DKR. Ind. Moll. Mar. Jap. p. 156.

A thin species, always depressed, remarkable for the endless variety of its mottled coloring. It is allied to *P. amussitata*, but the delicate riblets or striæ are not beaded as in that species.

P. GRATA Gould. *Unfigured.*

Shell ovate-conic, elevated, apex acute, very much anterior ; outside rude, ashen, with elevated compressed radiating ribs which are tubulose toward the margin ; margin expanded, denticulate. Inside ochraceous variegated with brown, spatula and submargin intense chestnut. Length 30, width 24, alt. 14 mill. (*Gld.*)

North shores of Niphon.

Patella grata GLD., Proc. Bost. Soc. N. H. vii, p. 161 (Dec., 1859); Otia Conchologica, p. 115.

This may be an *Acmæa*. It has not been figured, and I have not seen it.

(4.) NEW ZEALAND AND AUSTRALIA SPECIES.

The Patellas of New Zealand are all, with the exception of *P. tramoserica*, confined to that province. They have been referred by Hutton to *Patinella*, but erroneously, the branchial cordon being interrupted in front as in all *Helcioniscus*, a group with which they agree in dentition as well. The correct synonymy of many of the species is here given for the first time.

The dentition of several species has been figured by Hutton, Trans. N. Z. Institute, xv. A useful paper on the anatomy of *P. radians* has been published by J. A. Newell, Trans. N. Z. Inst. xix, p. 157, plate xi, 1887.

H. REDIMICULUM Reeve. Pl. 23, figs. 1, 2, 3, 5.

Shell oblong, rather depressed, solid, the apex between the front fourth and sixth of the shell's length, and inclined forward. Sculptured with about 22 rounded ribs. Ribs dark or buff, intervals bluish-white; having several darker concentric streaks, and marked near the apex with oblique black stripes.

Interior somewhat iridescent, obscurely rayed, having a cream-white central callus, often more or less bordered behind with olive; muscle-scar slightly pinkish.

Length 41, breadth 32, alt. 12 mill.

Length 43, breadth 34, alt. 12½ mill.

Southern New Zealand and Auckland Is.

P. redimiculum REEVE, Conch. Icon., f. 50, (1854).—HUTTON, Cat. Mar. Moll. N. Z. 1880, p. 107.—E. A. SMITH, Voy. Erebus & Terror, Moll., p. 4, t. 1, f. 24.—*P. radians Gm.*, REEVE, Conch. Icon., f. 25, (not of Gmelin).—*P. pottsi* HUTTON, Cat. Mar. Moll. N. Z. 1873, p. 44, *teste* Hutton.—*Patinella redimiculum Rv.* HUTTON, Proc. Linn. Soc. N. S. Wales ix, p. 375, (1884.)

The rounded ribs are nearly smooth. The coloring of oblique blackish stripes around the anteriorly curved apex is characteristic, but shared by some other species.

H. STRIGILIS Hombron & Jacquinot. *Unfigured.*

Shell oval, convex, obliquely conical; two-colored outside, above blackish-rufescent, below brownish-rufescent, having few white dots; principal radiating ribs 24, subequal, obtuse, subprojecting beyond the margin; vertex obtuse, white, excentric. Interior blackish-purple, the depth pale yellowish. Length 65, breadth 50 mill. (*H. & J.*)

Auckland Is. (H. & J.); *Banks' Peninsula to Shag Point, Otago, New Zealand; Auckland Is.; Campbell Id.* (Hutton).

P. strigilis H. & J., Ann. Sci. Nat. (2) xvi, p. 190 (1841).—*P. magellanica* HUTTON, Trans. N. Z. Institute xv, t. 16, f. A (dentition only); Man. N. Z. Moll., p. 107 (1880), not of Gmelin!—*Patinella strigilis* HUTTON, Proc. Linn. Soc. N. S. Wales, ix, p. 374 (1884).

The original description is given above. This species I have not seen. It certainly has nothing to do with *Patinella magellanica, ænea, kerguelensis* or *fuegiensis.*

Hutton's description in P. L. S. N. S. W., 1884, is as follows:

"Shell large, solid, obliquely conical, high, with about 20–30 low radiating ribs; the apex subcentral or rather anterior. Brown, obscurely marked with yellowish; interior greenish or yellowish-brown above the muscle impression, bluish-white and iridescent below it, the margin brown."

The description in Man. N. Z. Moll. does not, of course, apply to this shell.

H. ORNATUS Dillwyn. Pl. 68, figs. 14, 15, 16, 17, 18, 19; pl. 19, figs. 39, 40.

Shell solid, oval or oblong, rather low-conical, the apex at about the front third, erect. Surface having larger radiating, *coarsely nodular* ribs, about 11 in number, with a somewhat smaller rib between each pair of larger ones, the intervals radiately striated; growth-striæ fine, often quite distinctly cutting the radial striæ. *The larger ribs are light, the intermediate ribs are black dotted with white,* especially in the young, this coloring being less obvious on large shells.

Interior having alternating silvery and black rays, the latter usually 11 in number; the large central area black, suffused more or less with cream color in the depth of the apex.

Length 32, breadth 25, alt. 10 mill.

Throughout New Zealand.

Patella ornata DILLWYN, Descriptive Catal. Recent Shells, ii, p. 1029, (1817)).—*P. nodosa* HOMBRON & JACQUINOT, Ann. Sci. Nat. (2), xvi, p. 191, (1841).—*Patella margaritaria, testa ovali, etc.*, CHEMNITZ, Conchyl. Cab. xi, p. 180, t. 197, f. 1914, 1915 (1795).— *P. denticulata* E. A. SMITH, Voy. Erebus & Terror, Zool., ii, Moll., p. 4, t. 1, f. 26.—*P. luctuosa* GLD., Proc. Bost. Soc. N. H. ii, p. 150, (1846); U. S. Expl. Exped. Moll. & Sh., p. 336, f. 446.—*Patinella denticulata* HUTTON, Proc. Linn. Soc. N. S. Wales, ix, p. 375; Trans. N. Z. Inst. xv, t. 16, f. B (dentition).—*P. inconspicua* GRAY, in Dieffenbach's N. Z. ii, p. 244, (1843).—*Patinella inconspicua* HUTTON, Proc. Linn. Soc. N. S. W. ix, p. 375.—*Patella margaritaria* REEVE, Conch. Icon., f. 74.

Easily distinguished by its coarsely nodose ribs, eleven-rayed interior with black central area, and in the young by the alternately white-spotted riblets.

It is a very distinct species, but it has been afflicted with a number of names unusual even in this genus. It has no especially close relations with *P. denticulata* Martyn (*q. v.*)

Reeve's figures of *margaritaria* are copied on pl. 19, figs. 39, 40. They represent a large, rather round specimen.

The correct synonymy is herein given for the first time. The name *margaritaria* Chemnitz cannot be used, as that author was not a binomialist.

Var. INCONSPICUA (Gray) Hutton.   Pl. 68, figs. 20, 21, 22.

Shell conical, high, the altitude often more than half the length; apex subcentral. Interior brown, with about twelve radiating white stripes. (*Hutton.*)

Wellington to Dunedin, New Zealand.

The figures represent the *luctuosa* Gld. (not Hombr. & Jacq.) which is the same as *inconspicua.* Gray's description is very poor.

P. DENTICULATA Martyn.   Pl. 68, figs. 23, 24; pl. 21, figs. 49, 50.

Shell solid, oval, elevated, the apex more or less anterior; sculptured with about 31 principal ribs, and some smaller interstitial riblets *all of them closely scale-granose.* Color blackish-brown.

Interior bluish, having a distinct, opaque, flesh-colored or dull orange-brown central area. Spotted with brown around the edge.

Length 55, breadth 43, alt. 24 mill.

Wellington to Dunedin, New Zealand.

Patella denticulata MARTYN, Univ. Conch. i, t. 65 (1784).—*P. imbricata* REEVE, Conch. Icon. f. 95 (1855), not *P. imbricata* Linné. —*P. reevei* HUTTON, Man. N. Z. Moll. p. 108 (1880).—*Patinella reevei* HUTTON, Proc. Linn. Soc. N. S. Wales, ix, p. 376 (1884).

Differs from other New Zealand species in the closely scaled ribs, opaque "rust orange" or fleshy central area of the interior, etc. It is a solid species, attaining a considerable size.

Reeve referred *denticulata* Martyn to *P. granularis* as a synonym and made a new species, "*imbricata*" of the New Zealand shell. Hutton has followed Smith in mistaking the *P. ornata* Dillw. (*margaritaria* Chemn.) for Martyn's shell—an error immediately detected by a reference to the excellent figures in the Universal Conchologist.

In order to finally settle the name and synonymy of this species, I have copied Martyn's original figures on my plate 68, figs. 23, 24.

P. RADIANS Gmelin. Pl. 69, figs. 25–39; pl. 23, figs. 4, 6, 7, 8.

Shell ovate, depressed, thin but solid, slightly narrower in front, the apex at the front fourth or fifth, not prominent. Surface sculptured with decidedly *separated, narrow radiating riblets*, having a number of smaller riblets (sometimes obsolete), in each interval, and decussated by fine, crowded growth-striæ, often obsolete, but usually *cutting the surface just in front of the apex into fine granules*. Color bluish-white, usually buff around the apex, *striped in a divaricating pattern*, or irregularly blotched and rayed down the ribs with brown or olive.

Interior buffish-olive, with a silvery luster, showing the color-markings of the outside, having a white or brown central callus, often ill-defined.

Length 44, breadth 34, alt. 8 mill. (typical form).

Throughout New Zealand; Australia.

P. radians GMEL., Syst. Nat. xiii, p. 3720 (1789).—HUTTON, Man. Mar. Moll. N. Z. 1880, p. 108; Trans. N. Z. Inst. xv, t. 16, f. E (dentition).—*P. argentea* Q. & G., Voy. Astrol. Zool. iii, p. 345, t. 70, f. 16, 17 (1834).—*P. argyropsis* LESSON, Voy. Coquille, p. 419 (1830).—*P. radiatilis* HOMBR. & JACQ., Ann. des Sci. Nat. (2), xvi, p. 191 (1841).—*P. decora* PHIL., Zeitschr. f. Mal. 1848, p. 162; Abbild. t. 3, f. 3.—RVE., Conch. Icon. f. 33.—*Patinella radians* HUTTON, Proc. Linn. Soc. N. S. Wales, ix, p. 336 (1884).—*Patella earlii* REEVE, Conch. Icon. f. 71 (1855).—*P. flexuosa* HUTTON, Cat.

Mar. Moll. N. Z. p. 45 (1873), not of Q. & G.—*Lottia radians* Sow-
ERBY (*de novo*), Genera, Lottia f. 3.—*Patella affinis* REEVE, Conch.
Icon. f. 108, (1855.)—*P. fusca* LINN., Syst. Nat. x, p. 784.—HANLEY,
Shells of Linn., p. 428, t. 4, f. 9 (fig. of Linnæus' type specimen).—
P. sagittata DONOVAN, Rees' Encyclop., Conchol. t. xvi.

The earliest name proposed for this shell is that of Linné, *P. fusca.*
It was defined in an absurdly inadequate manner, however, and as
Hanley justly remarks, no claims to precedence can be grounded
upon the mere preservation of the original specimens.

Typically depressed, but sometimes as elevated as *P. redimiculum.*
The sculpture of *narrow separated riblets*, having in the intervals a
smaller riblet, or numerous minute riblets (sometimes obsolete), is
characteristic. Sometimes the whole surface between the larger
ribs is finely granulose, and most specimens retain this granulation
in front of the apex.

The connecting forms are so numerous that I am unable to
diagnose any of the following as varieties worth naming.

Typical form (pl. 69 figs. 25–28). Much depressed, thin, riblets
often subobsolete ; conspicuously striped and blotched with brown or
red on a bluish-white ground, yellow around the apex.

The form called *argentea* by Q. & G. is dark, mostly olivaceous,
depressed.

The *P. decora* of Philippi (pl. 69, figs. 29–31) is yellowish with
about 24 narrow, reddish-brown ribs, alternating with small riblets.
Reeve's *decora* is intermediate between this and *argentea* Q.

Reeve's *P. earlii* (pl. 21, figs. 51, 52) is typically more elevated,
rounder, "pale green, broadly wave-variegated with olive-black."
It is very closely connected by intervening forms with the type.

The *P. affinis* of Reeve (pl. 69, figs. 32, 33) scarcely differs from
the typical *radians.* "The surface is carved throughout with simple
smooth, slightly waved, close-set ridges and striæ."

Var. PHOLIDOTA Lesson. Pl. 69, figs. 38, 39.

Ribs small and uniform ; apex very anterior, about one-seventh
of the length from the anterior end. Olive-brown, largely blotched
with white, or white with brown radiating bands. (*Hutton.*)

Throughout *New Zealand.*

P. pholidota LESSON, Voy. de la Coquille, p. 420 (1830).—*P.
sturnus* HOMBR. & JACQ., Ann. des Sci. Nat. (2), xvi, p. 191, (1841).

P. floccata Reeve, Conch. Icon., f. 106, (1855).—*P. radians var. pholidota* Hutton, Proc. Linn. Soc. N. S. Wales, ix, p. 377, (1884).

The figures represent the synonymous *P. floccata* of Reeve.

H. OLIVACEUS Hutton. Pl. 70, figs. 46, 47, 48.

Shell short-ovate, conical, the apex at about the front third; closely but subobsoletely radiately ribbed, the riblets about 70 in number, and of nearly equal size; uniform olive colored.

Interior greenish-olivaceous, iridescent, the center whitish; edge narrowly black-bordered.

Length 33, breadth 28, alt. 14 mill.

Dunedin to the Bluff, New Zealand.

P. olivacea Hutton, Trans. N. Z. Inst. xv, p. 133, (1883), pl. 16, f. D (dentition).—*Patinella radians var. olivacea* Hutton, Proc. Linn. Soc. N. S. Wales, ix, p. 377, (1884.)

The apex is less anterior than is usually the case with *P. radians,* and it is more erect. It is further distinguished by the uniform olive color, and more equal ribbing. My illustrations are from a specimen received from Prof. Hutton.

This species seems quite distinct from the *P. radians,* of which I have seen many examples. Prof. Hutton, however, considers it a variety of that species, no doubt having good reasons for the union.

There is a prior *P. olivacea* of Anton, and a still earlier *P. olivacea* of Gmelin, but as both are totally unrecognizable, the name imposed by Hutton may be allowed to stand.

H. STELLIFERA Gmelin. Pl. 70, figs. 43, 44, 45.

Shell depressed, oval, with small granular ribs; reddish with white rays at the apex, or two white lines at the posterior end. Interior white. Apex anterior. Length 25, breadth 19, alt. 7 mill. (*Q. & G.*)

Cooks' Straits to Bank's Peninsula, New Zealand.

Patella stellata, seu stellifera, etc. Chemnitz, Conchyl. Cab. x, p. 329, f. 1617 (1788).—*P. stellifera* Gmelin, Syst. Nat. xiii, p. 3719. —*P. stellularia* Quoy & Gaim., Voy. Astrol. Zool. iii, p. 347, t. 70, f. 18–20 (1834).—Reeve, Conch. Icon. f. 96.—*Patinella stellifera* Hutton, Proc. Linn. Soc. N. S. Wales, ix, p. 378.

Readily recognized by the central white star.

H. TRAMOSERICA Martyn.   Pl. 70, figs. 49, 50, 51, 52.

Shell solid, short-oval, conical, the apex erect, near the center or somewhat anterior.   Surface sculptured with numerous (about 40) narrow ribs, with usually an interstitial small riblet in each interval, the concentric striæ of growth crowded, sometimes prominent enough to finely crenulate the radiating ribs.   Color varying from yellowish with blackish-brown rays, to reddish-brown with whitish rays.

Interior yellowish, lustrous, having dark rays and spots; central area having a whitish, orange or olive callus.

Length 46, breadth 40, alt. 17 mill.

Wellington, New Zealand; Chatham Is.; New South Wales, Australia.

P. tramoserica MARTYN, Univ. Conch. i, t. 16.—REEVE, Conch. Icon. f. 27.—*P. antipodum* E. A. SMITH, Voy. Erebus and Terror, Moll. p. 4, t. 1, f. 25 (1874).—*Patinella tramoserica* HUTTON, Proc. Linn. Soc. N. S. Wales, ix, p. 377 (1884).—? *P. pecten* GMEL., Syst. xiii, p. 3702.

The ribs are numerous, their narrow intervals usually having an interstitial riblet.   The color is dull yellow or reddish, with dark rays, which are sometimes seen to be fretted or dotted if held toward the light.

H. FLAVUS Hutton.

Ovate, conical, radiately ribbed; apex recurved; margin crenated; pale yellow, inclining to orange toward the apex; interior, above the muscular-impression more or less orange, below silvery.

Length 2·2, breadth 1·8, alt. 1 inch.   (*Hutton.*)

Poverty Bay to Stronghurst, Canterbury, New Zealand.

P. flava HUTTON, Cat. Mar. Moll. N. Z., p. 44, (1873); Man. N. Z. Moll., p. 109, (1880).—*Patinella flava* HUTTON, Proc. Linn. Soc. N. S. Wales, ix, p. 378.

The description of this form is scarcely sufficient.

H. ILLUMINATA Gould.   Pl. 70, figs. 40, 41, 42.

Shell elevated conical, with an arched outline, the apex at about the anterior fourth; surface covered with numerous small, obtuse, radiating ribs, with from one to three intervening striæ; concentric lines of growth crowded, very faint.   Color sooty, with scattered, yellowish spots, about twenty in number somewhat regularly dis-

posed, which are transparent when held up to the light, those near
the margin elongated. Aperture ovate, the margin slightly irregular; interior a very dark claret-color, with brilliant silky and
golden reflections, and yellow spots, corresponding to those of the exterior; central spatula dull buff-color. (*Gld.*)

Length 1⅓, breadth 1⅛, alt. ⅜ inch.

Auckland Is. (Gld.); *Campbell and Macquarie Is.* (Hutton.)
P. illuminata GLD., Proc. Bost. Soc. N. H. ii, p. 149, (1846);
Exped. Moll. & Sh., p. 340, atlas f. 441.—HUTTON, Trans. N. Z.
Inst. xv, t. 16, f. c (dentition).—*Patinella illuminata* HUTTON, Proc.
Linn. Soc. N. S. Wales, ix, p. 376.—? *Patella terroris* FILHOL, Compt.
Rend. xci, (1880.)

* * *

Australian species.

H. LIMBATA Philippi.   Pl. 71, figs. 53, 54, 55, 56; pl. 17, figs. 28,
29.

Shell solid, oval, the apex near the front third; sculptured with
from 21 to 31 *broad rounded ribs.* Brown or reddish, *the intervals
between the ribs brown-striped.* Apex eroded. Edge scalloped by
the ribs.

Inside whitish, tinted with flesh-color or lilac, the central callus
generally opaque-white mixed with bluish, becoming olive toward
its edges. There is a distinct yellow or brown border, within which
the stripes of the exterior make a ring of vivid blotches, these stripes
being also, at times, visible through the tinted lining of the shell.

Length 47, breadth 41, alt. 20 mill.

Length 59, breadth 52, alt. 29 mill.

Port Lincoln, S. Australia; Tasmania.

P. limbata PHIL., Abbild. u. Beschreib., iii, p. 71, *Patella* t. 3, f.
2.—REEVE, Conch. Icon. f. 29.—ANGAS, P. Z. S. 1865, p. 185.—
TENISON-WOODS, Proc. Roy. Soc. Tasm. 1876, p. 48 (animal).

A large, solid species. The broad, rounded ribs, narrow interstices striped with brown, and distinct internal border are characteristic marks of this species. The eroded apex is usually stained
with bluish. The number of ribs is excessively variable, sometimes
as many as 37 being developed.

H. LATISTRIGATA Angas.

Shell very similar to *P. limbata*, but more elongated; liver-colored,
rayed with a few very broad brown-black stripes. Ornamented with

about 1? to 14 irregular, rounded ribs. Spatula of an intense brown-black, margined with white. (*Angas.*)

Length ·7, breadth ·45, alt. ·2 inch.

Aldinga Bay, South Australia.

P. latistrigata ANGAS, P. Z. S. 1865, p. 154, 186.

H. GEALEI Angas.

Similar to *P. Jacksoniensis*, but the interior splendidly metallic, a little tinted with golden; margin narrow. Spatula lurid, clouded with leaden and brown. (*Angas.*)

Length ·1, breadth ·86, alt. ·4 inch.

St. Vincent's Gulf, S. Australia.

P. gealei ANG., P. Z. S. 1865, pp. 57, 186.

H. ARANEOSA Reeve. Pl. 71, figs. 57, 58.

Shell suboblong-ovate, attenuated in front, rather thin, compressed at the sides, apex rather sharply acuminated, anterior; radiately densely striated, striæ corded, minutely crenulated with concentric striæ. Olive-green, conspicuously rayed with numerous opaque-white lines. Interior semitransparent.

A somewhat depressed species, pinched and sharply pointed toward the apex, which inclines very much to the front, and of a dull greenish-olive color, curiously rayed throughout with fine opaque-white lines. (*Rve.*)

Australia.

P. araneosa REEVE, Conch. Icon., f. 111. March, 1855.

Gould described in 1846 a different species under this name, but it is probably an *Acmæa.* See appendix to this volume.

(5.) SPECIES OF SOUTH AND EAST AFRICA AND ADJACENT ISLANDS.

Limpets of the genus *Helcioniscus* are not found on the West African coast, which is inhabited by the typical groups of *Patella.* Upon the East African shores, however, typical *Patella* is not numerous in species, but *Helcioniscus* abounds from the Red Sea to the Cape.

H. ROTA Gmelin. Pl. 72, figs. 65–80.

Shell oval, low-conical, the apex at the front third or behind it. Sculptured with numerous subequal or unequal, obsoletely granulous radiating striæ. Outside whitish, having purple-brown rays which *sometimes branch to form v's, sometimes are spotted with light* or *split into several narrow stripes.* The rays are generally eleven in number, but often some of them are multiplied by splitting.

Interior yellow (or silvery), showing the rays as vividly as the outside; the central area red-chestnut in color.

Suez and *Mozambique* (Reeve); *Madagascar* (Dall); *Reunion* (Nevill, in Phil. Acad. Colln.).

Patella rota, testa subrotunda, etc., CHEMN., Conchyl. Cab. x, p. 330, t. 168, f. 1619.—*P. rota* GMEL., Syst. Nat. xiii, p. 3720.— REEVE, Conch. Icon. f. 39a.—*Helcioniscus rota* DALL, Amer. Jour. Conch. vi, p. 278, pl. 16, f. 28.—*Patella variegata* REEVE, Conch. Syst. t. 136, f. 1.—*P. petalata* RVE., Conch. Icon. f. 56.—*P. luzonica* RVE., *l. c.,* f. 86.—*P. scalata* RVE., *l. c.,* f. 89.—?? *P. argentaurum* LESSON, Voy. de la Coquille, p. 414.

An excessively variable species. It is closely allied to *P. novem-radiata,* but in that species the rays do not bifurcate, although they are often split into two, and the indistinctly defined central callus is white with more or less bright yellow tint. It is even more closely allied to *P. capensis* Gmel and *P. variabilis* Krauss, and the three may prove to be one species, although in some details they differ. I have seen too few of the Cape species to write confidently upon their constant characteristics.

Var. ROTA, typical, (figs. 65–69) may be restricted to the forms with chestnut or reddish spatula. As a synonym I have placed *P. variegata* Reeve, founded upon a typical specimen of *rota,* although Reeve afterward shifted the name to a different species.

The following seem to belong here:

P. petalata Rve. (pl. 72, figs. 70, 71). "Shell ovate, rather depressed, obsoletely cancellately ridged; transparent yellow, painted with broad blackish-purple rays. Interior transparent horny, nucleus chestnut-purple. *Australia.*" (*Reeve.*)

P. luzonica Reeve, (pl. 72, figs. 72, 73). "Shell rotundately ovate, rather depressed, radiately finely ridged, ridges granuled; apex sharp, anterior minutely hooked; transparent yellow, rather horny, promiscuously stained with large black blotches. Interior transparent, subiridescent, dark chestnut in the middle. *Luzon, Philippines.*" (*Rve.*)

P. scalata Rve. (pl. 72, figs. 74, 75). "Shell ovate, rather sharply convex; apex scarcely central; radiately obtusely striated, and here and there linearly grooved; livid-white, rayed with black
10

bands peculiarly bi-forked, or diagonally linearly streaked. Interior subtransparent, horny. *Philippine Is.*" (*Rve.*)

Var. *orientalis* Pilsbry. Pl. 72, figs. 76, 77.

Shell more solid, the central area of the interior whitish or more or less stained with olive or orange.

Viti Islands (A. Garrett).

Var. DISCREPANS Pilsbry. Pl. 72, figs. 78, 79, 80.

Surface with growth-lines but no radiating sculpture whatever. Soiled white, with purple-brown rays torn into oblique shreds. Length 29, breadth 24, alt. 10 mill.

Probably a distinct species. The two specimens I have seen are unlike any described *Helcioniscus* in their smoothness and lacerated rays.

H. CAPENSIS Gmelin. Pl. 16, figs. 15, 16, 17.

Shell ovate, thin, depressed-conical; dull white, variously painted with brown radiating bands and spots ; radiately striated, the striæ close, equal, *granulose ;* vertex acute, erect, situated at about ⅓ the length ; margin denticulated.

Interior *yellowish-silvery, with a pearly* luster, having rays and spots of brown ; central spatula, *brown or orange,* rarely whitish, but *always marked with white under the vertex and a brown spot in front.* (*Krauss.*)

Length 39, breadth 29 mill.

Natal.

P. capensis GMEL., Syst., p. 3720.—KRAUSS, Die Südafric. Moll. p. 53, t. 3, f. 13.

There is a white area in the depth of the interior, having an orange or brown bar across it, as in pl. 16, figs. 16, 17 ; or the white is reduced to a bar in the same place.

Compare *P. rota* Gmel.

H. NOVEMRADIATUS Quoy & Gaimard. Pl. 30, figs. 55, 56, 57, 58.

Shell low-conic, *rounded-oval,* rather thin but solid, the apex slightly in front of the center. Surface lusterless, *closely, finely striated* radially, the striæ somewhat granulose, often subobsolete ; growth-lines obvious or obsolete. Whitish, broadly rayed with olive-brown or dull rust-red.

Interior layer *translucent, iridescent, conspicuously showing the rays of the outside,* which become vivid brown at the edge ; central area having an ill-defined callus, which is more or less deeply *stained with bright gamboge yellow.* Length 40, breadth 34, alt. 10 mill.

Mauritius.

P. novemradiata Q. & G., Voy. de l'Astrol. p. 346, t. 70, f. 22, 23. —*P. aster* REEVE, Conch. Icon. f. 80, 1855.

A splendid species. The rays are broad and about nine in number, but more frequently they are twice as numerous by the splitting of each broad one into two. Quoy & Gaimard described a very young shell. Reeve figured under the name *P. aster*, a small specimen from an unknown locality. The series before me is from Mauritius (Robillard Coll.).

It is a much larger, more spreading and vividly colored shell than the allied *P. profunda.* There is little besides coloring to sunder this species from *P. rota* Gm. Compare also *P. capensis* and *P. variabilis*, the former of which may be the same. It should be noted that Gmelin refers to Kaemmerer, t. 2, f. 1, 2, as an illustration of his *capensis.* These figures represent the typical *novemradiata.*

H. VARIABILIS Krauss. Pl. 16, figs. 18, 19, 20.

Shell ovate, thin, depressed-convex; whitish or ashen-yellow, painted with radiating bands and spots of ashen or brown; radiately striated, striæ or riblets unequal, *transversely very minutely striated;* vertex acute, looking forward, situated at the front third; margin denticulated.

Interior yellowish, rarely whitish, having radiating bands and spots of brown, *shining.* Central area not distinct, yellowish or whitish. (*Krauss.*)

Length 32, breadth 24, alt. 7–9 mill.

Natal.

P. variabilis KRAUSS, Die Südafric. Moll., p. 35, t. 3, f. 12 (not *P. variabilis* Sowb., a species of *Acmœa.*)

This is a flatter shell than the preceding, rather thin, translucent, rarely eroded at the apex. The apex is more forwardly directed than in *P. capensis.* From it many (70–80) alternately larger striæ radiate, which are cut by very fine concentric striæ, scarcely visible with the naked eye. The color outside is usually dirty white or yellowish with many grayish-brown or brown radiating striæ or flecks, which are visible with more intense color through the grayish-yellow or whitish-yellow shining (but never silvery and pearly) layer of the interior.   The centrum is not sharply defined, generally scarcely darker than the circumference.   (*Krauss.*)

Compare *P. rota* Gmel.

The following variations are described by Krauss:

Var. *fasciata* (fig. 18).   Shell whitish, spotted with brown, painted with six broad blackish-brown bands.   The typical form is irregularly and interruptedly striated and flecked, but this has broad bands.

Var. *radiata* (fig. 19).   The usually somewhat stronger riblets are white, and the smaller riblets and grooves are brown or blackish-brown.   Under a lens young examples are seen to be sprinkled with little light-blue flecks.   The centrum is yellowish.

Var. *concolor*.   Shell unicolored, blackish-ashen or tawny.   Always smaller, totally unicolored.   Centrum whitish.

H. DUNKERI Krauss.   Pl. 16, figs. 11, 12, 13, 14.

Shell small, ovate, convex, very thin; subpellucid; whitish or dull yellowish, having 11 radiating black bands and reddish striæ, sometimes painted with rose and spots of bluish-green; radiately striated, the striæ fine, subequal; vertex acute, inclined forward, situated at the front fourth; margin very finely denticulated, not gaping.   Interior shining, colored like the outside, the center yellowish or ashen-whitish.   (*Kr.*)

Length 17, breadth 11, alt. 4¾ mill.

Natal.

P. dunkeri KR., Die Südafric. Moll. p. 55, t. 3, f. 14.—REEVE, Conch. Icon. f. 124.—PHIL., Abbild. t. 2, f. 9.

The thinnest of the South African species.   It is somewhat intermediate between *P. variabilis* and *P. pruinosa*, according to Krauss.   Compare also *P. compressa* young, and *P. araneosa* Rve.

H. EUCOSMIA Pilsbry.   Pl. 71, figs. 61, 62, 63, 64.

Shell oval, conical, the distance in front of the apex contained from 2½ to 3 times in the length of the shell.   Posterior slope somewhat convex.   Sculptured with fine closely granulous radiating riblets, of which every fourth one is usually somewhat larger.   Outside gray-white, spotted all over and indistinctly rayed with rusty-brown.

Interior yellowish, *conspicuously blotched, spotted and rayed* with purple-brown or black-brown; the rays being usually 11 or 12 in number, either wide or narrow, and spotted with light.   The central callus is dark orange-brown, sometimes encircled by a whitish ring.

Length 40, breadth 32, alt. 15 mill.

Suez (Fischer); *Red Sea and Gulf of Akaba* (Smithsonian Cabinet): *Japan at Hakodadi* (Stimpson and Anthony in Phil. Acad. Colln.); *Australia.* (Rve.)

P. variegata REEVE, Conch. Icon. species 38 (Dec. 1854), not *P. variegata* REEVE, Conch. Syst. pl. 136, fig. 1 (1842).—*P. variegata* FISCHER, Journ. de Conchyl. 1870, p. 167.—*Helcioniscus variegatus* DALL, Amer. Journ. Conch. vi. p. 277, t. 16, f. 27 (animal and dentition).—Not *Patella variegata* DE BLAINVILLE, Dict. des Sci. Nat. xxxviii, p. 100 (1825).

Readily distinguished from *P. rota* by its blotched and speckled color-pattern.

Part of the localities given above are no doubt incorrect. I have examined a very large series, and find but little variation from the typical form. I have seen no specimens approaching *P. rota.*

Reeve in 1842 described and figured a specimen of typical *P. rota* Gm. under the name *variegata.* In 1854 he shifted that name to the present species, giving no reason for such change, nor even admitting that he had made a change. Under these circumstances it becomes necessary to give a new name to the present species, and thus avoid the confusion otherwise inevitable. It should also be noted that there is a prior *P. variegata* of Blainville, 1825.

H. CERNICA (Barclay) Adams.   Pl. 71, figs. 59, 60.

Shell thin, ovate, depressed-conic; decussated with numerous obtuse radiating ribs and close, elevated, undulating concentric liræ; whitish, ornamented with reddish-brown rays. Apex subcentral, obtuse; aperture ovate.

Interior sculptured and colored like the outside, shining, somewhat pearly; margin more or less widely crenulated.

Length 39, breadth 29, alt. 10 mill. (*Ad.*)

Barkly Island, Mauritius.

Nacella (Cellana) cernica Barcl. *ms.*, H. ADAMS, P. Z. S. 1869, p. 273, t. 19, f. 7, 7a.

This species is the type of H. Adams' subgenus CELLANA. It probably belongs to *Helcioniscus* rather than to *Nacella* or *Patinella.*

The name *Cellana* has priority over *Helcioniscus* but it has not been adequately defined.

II. PROFUNDUS Deshayes. Pl. 65, figs. 94, 95, 96.

Shell small, elevated-conical, solid, apex a little anterior, the slopes straight; *surface finely closely and evenly radiately striated*, white with purplish or brownish rays; the rays usually 10 in number, articulated with darker spots which are often angular.

Inside white, showing the rays faintly; the central area light chestnut or outlined with light chestnut, edge of the shell smooth, articulated with white and chestnut.

Length 16, breadth 12, alt. 7 mill.

Length 20, breadth 15½, alt. 8 mill.

Island of Réunion.

P. profunda DH., Moll. Réunion, p. 44, t. 6, f. 15, 16, 1863.

A small conical species, nearly smooth, the radiating striæ being quite fine. The rays are more obvious on worn examples. The central area is not calloused. It has very much the appearance of an *Acmæa.*

Var. MAURITIANA Pilsbry. Pl. 65, figs. 97, 98, 99.

Shell thicker, heavier, more elevated; dull white with reddish rays, which are not visible within, and only faintly visible at the edge, which is minutely crenulated; basal side-margins slightly curving upward; radiating striæ of the surface coarse, unequal. Central area of the interior having an orange-tawny callus.

Length 21, breadth 17, alt. 11 mill.

Mauritius.

H. GARCONI Deshayes. Pl. 66, figs. 100, 101.

Shell ovate, little narrowed in front, conical; vertex acute, subcentral; ornamented with small granulous radiating striæ; blackishbrown, vividly pearly inside, toward the apex whitish. (*Dh.*)

Regularly oval, conoidal, the summit elevated, pointed, very slightly directed forward, situated at the front two-fifths of the length. From the apex radiate a great number of very fine, regular, rather equal riblets, which bear long, obtuse granules. The margins are simple and sharp. The interior is lined with very bright nacre of a whitish-brown, the central callus quite large, white, sharply defined by the muscle-scar. The shell is thin, semitransparent, of a uniform brown-blackish, but if held up toward the light, a few rays of a beautiful red become visible. (*Dh.*)

Length 23, breadth 19, alt. 9 mill.

Island of Réunion.

P. garconi Dh., Moll. de l'Ile Réunion, p. 42, t. vi (xxxiii), f. 11, 12.

I have not seen this species, which apparently resembles *P. olivacea* and *P. ardosiæa;* but those species are truly unicolored, whilst this shows rays when held toward the light. It is probably allied to *P. profunda* Dh.

H. DEPSTA Reeve. Pl. 20, figs. 45, 46.

Shell ovate, thin, rather depressed, raised in the middle, rather compressed at the sides; apex sharp, anteriorly minutely hooked; radiately striated, striæ raised toward the margin; reddish-chestnut, sometimes faintly rayed with greenish-yellow. Interior livid horny.

A nearly smooth reddish-chestnut shell, with a sharp minutely hooked apex. (*Rve.*)

Macao and Island of St. Paul.

P. depsta REEVE, Conch. Icon., f. 85, 1855.

H. SANGUINANS Reeve. Pl. 30, figs. 53, 54.

Shell oblong-ovate, rather convex, apex rather anterior; decussated with concentric striæ and small superficial ridges; whitish, here and there peculiarly rough, marked with promiscuously flowing blood-red streaks, rough surface red-dotted. Interior semipellucid white, conspicuously red rayed.

A fine new species of an oblong-oval form, rather flattish, with the apex situated somewhat anteriorly. It is of a reddish-white ground, painted with promiscuous streaks of red, like blood flowing, and the ground is singularly overlaid here and there with an opaque, rough coating dotted with red. (*Rve.*)

Cape Natal, S. Africa.

P. sanguinans RVE., Conch. Icon., f. 10a, 10b. Oct., 1854.

I have not seen this species. It is not mentioned by Krauss or other writers on the Cape fauna.

(6.) SPECIES OF UNKNOWN HABITAT.

H. MELANOSTOMUS Pilsbry. Pl. 32, figs. 67, 68, 69.

Shell solid, erectly elevated-conical, the base ovate; slopes nearly straight; apex subcentral, erect. Surface sculptured with numer-

ous (43–45) strong rounded ribs, closely but usually rather super-
ficially cut by concentric striæ. Color clear buff, unicolored or having the ribs black or black-spotted.

Interior white and silvery, the central callus of a more or less intense purple-black.

Length 61, breadth 51, alt. 34 mill.

Length 50, breadth 41, alt. 30 mill.

Habitat unknown.

Distinguished at once by the unusual coloring of the interior,
and the sculpture of the outside. All of the specimens before me
have the upper portion of the cone eroded. In one the central
black callus is very thick. The ribs are alternately larger and
smaller. They are scarcely strongly enough represented in the figure.

Specimens having black ribs possess also two short dark streaks
radiating from the forward angles of the "head-mark" as in *P.
boninensis,* with which species this is probably most nearly allied.

H. ENNEAGONA Reeve. Pl. 28, figs. 35, 36.

Shell ovate, a little attenuated in front, depressed, rather thick,
obtuse at the apex, subattenuated and inclined, densely crenulately
ridged, peculiarly nine-sided, the three front areas narrow. Yellowish, the nine areas diagonally reticulately streaked with deep purple.
Interior silvery, iridescent. (*Reeve.*)

Habitat unknown.

P. enneagona REEVE, Conch. Icon., f. 44. (Dec., 1854.)

H. LIVESCENS Reeve. Pl. 73, figs. 99, 100.

Shell ovate, rather thin, depressly convex, rather sharply pointed
at the apex, radiately densely granulately striated, striæ slightly
waved; apex rather anterior. Pellucid blue-green, rayed with
blackish-purple, rays sometimes broken up into opaque blotches.
Interior iridescent-silvery, more or less transparent. (*Rve.*)

Mazatlan (Reeve.)

P. livescens RVE., Conch. Icon., f. 75, 1855.

A very delicate subpellucid greenish-blue shell, painted with dark
purple rays which show through into the interior. The sculpture
consists of numerous slightly waved granuled lines. (*Rve.*)

The locality given by Reeve is more than doubtful. It seems
near *P. rota* Gmel.

H. DIRUS Reeve. Pl. 73, figs. 88, 89.

Shell ovate, erectly conoid, densely radiately ridged and ribbed, ribs and ridges very closely broken up into small warts. Dark blue-black, sub-iridescent in the interior.

Although a similarity prevails between the figures of this and the preceding species [*P. guttata* Orb.], they are very different. *P. dira* being of a deep, erectly conical form, rayed throughout with close-set ribs and ridges, crenated with small tumid warts. (*Rve.*)

Habitat unknown.

P. dira REEVE, Conch. Icon. f. 92, 1855.

H. FUNGUS Reeve. Pl. 44, figs. 18, 19.

Shell ovate, depressly conoid, apex raised, subcentral; radiately densely ribbed and ridged, ribs and ridges narrow, small, everywhere finely noduled. Dull ash, dotted around the apex and near the margin with reddish-brown, with nodules whitish. Interior semitransparent, iridescent.

A dull ash, depressly conoid shell, densely rayed with finely noduled ribs and ridges, sparingly marked with reddish-brown dots, which are seen most distinctly in the interior of the shell. (*Rve.*)

West Indies (Rve.)

" *Tectura fungus* MEUSCHEN " *teste* Reeve.—*Patella fungus* RVE., Conch. Icon., f. 105, 1855.

This seems to be a *Helcioniscus* of the *H. exaratus* type. The locality given by Reeve is in all probability incorrect.

H. (?) ADELÆ Potiez & Michaud. Pl. 67, figs. 11, 12.

Shell ovate, depressed, blackish, painted with nearly regularly placed oblong white spots; radiately most finely costulated, the riblets very numerous, unequal, subgranulous. Vertex anterior, acute and tawny; margin acute, entire. Inside blackish in the depth of the cavity, the margin maculated, intermediate space whitish. Length 15, breadth 10, alt. 4 mill. (*P. & M.*)

Habitat unknown.

P. adelæ P. & M., Galerie des Moll., Mus. de Douai, i, p. 523, t. 37' f. 1, 2, 1838.

H. LINEATUS Lamarck. Pl. 73, figs. 85, 86, 87.

Shell oval, convex, buff-brown, painted with 10–12 yellow lines; excessively numerous longitudinal close striæ; vertex acute, buff. Length exceeding one inch. (*Lam.*)

Habitat unknown.

P. lineata LAM., An. s. Vert. vi, p. 331.—DELESSERT, Rec. de Coq. t. 28, f. 6.

I do not recognize Delessert's figures of the type of this species. Deshayes (in Lam. 2d. edit.) gives no additional information, as he did not have access to Lamarck's cabinet.

II. NIMBUS Reeve.   Pl. 35, figs. 30, 31.

I have described this under *Acmœa*, page 61, but it may be a *Helcioniscus*.   I have not seen the species.

Spurious, Unidentified and Unfigured Limpets.

The following list consists mainly of such species as I have been unable to identify with known forms, and which seem to me fairly unidentifiable.   A certain proportion of the number I have identified with certainty; and others with doubt, as will be seen by my notes below.

It will be understood that I have no desire to revivify names or species which have passed from the memory of man.   It is at least a half century too late for that.   It is not too late, however, to avoid the use of these dead names for new forms.   The frequent duplication of specific names in *Patella* has shown the necessity of a complete list such as that here given.   I have deemed it neither necessary nor desirable to change well-known modern names which are preoccupied by the old and insufficiently defined names given below, although there exist a considerable number of such duplications.

The species of *Clyptræidæ, Fissurellidæ, Siphonariidæ* and *Gadiniidæ*, etc., included by older writers in *Patella*, have been excluded, as far as the original descriptions have enabled me to judge of them.

? NACELLA SUBSPIRALIS Cpr. (Proc. Cal. Acad. Sci. iii, p. 213), belongs to the *Siphonariidæ*.

TECTURA RADIATA Pse. (Proc. Zool. Soc. Lond. 1860, p. 437), is a synonym of *Williamia gussoni* Costa—*Siphonariidæ*.

P. VIRIDIS Dufo (Ann. Sci. Nat. (2) xiv, p. 204, 1840).   *Seychelles and Amirantes.*   Unidentified.

P. AURIFERA Dufo (Ann. Sci. Nat. (2) xiv, p. 204).   *Mahé, Seychelles.*   Unidentified.

P. VIRGINUM Dufo (Ann. Sci. Nat. (2) xiv, p. 205).   *Mahé, Seychelles.*   Unidentified.

P. MALICOLOR Dufo (Ann. Sci. Nat. (2) xiv, p. 205). *Mahé Seychelles*. Unidentified.

P. DIEMENSIS Philippi.

Shell ovate-elliptical, convex conic, rather solid, whitish, having about 54 brown grooves; interior white, margin crenulated, the inside marked with brown dots at the crenations; apex at two-fifths of the length. Length 16½, breadth 14, alt. 8 lines. (*Phil.*)

Hobarttown, Tasmania.

P. diemensis PH., Zeitschr. f. Mal. 1848, p. 162.

Compare *Helcioniscus limbata* and *tramoserica*.

P. CRASSA Lesson (Voy. de la Coquille, Zool. ii, p. 413, 1830). A very thick, massive species, measuring, length 3⅘, breadth 3, alt. 1¾ inches.

Said to be from New South Wales.

P. COSTATA Lesson, (Voy. de la Coquille, Zool. ii, p. 415, 1830). The summit is said to be recurved posteriorly; sculptured with radiating striæ fine above, wide at the edge, surmounted by lamellæ. Outside green, surrounded with blackish toward the apex, having the elevations on the ribs bright chestnut. Inside golden red in the middle, silvery outside, rayed with blackish-purple at the positions of the ribs. Length 12, breadth 9, alt. 3 lines.

Island of Buru, Moluccas.

P. BOUROUNIENSIS Lesson, (Voy. de la Coq. Zool. ii, p. 415, 1830). Much depressed, oval, white, with black ribs outside, etc. Summit much posterior. Length 14, breadth 12, alt. 3½ lines.

Bay of Cajeli, Buru.

P. CROCATA Lesson (Voy. de la Coq. Zool. ii, p. 415, 1830). Irregularly oboval, little elevated, not thick, summit conical, nearly central, margins thin and angulose. Covered with close little ridges separated into little groups by shallow, spaced grooves. Color whitish with purple-violet rays. Inside white, tinted with yellow and silvery. Center saffron-yellow; border spotted with purple. Length 11, breadth 10, alt. 4 lines.

Port Praslin, New Ireland.

P. CONCEPSIONIS Lesson (Voy. de la Coq. Zool. ii, p. 418, 1830). Evidently an *Acmæa* or a *Scurria*, perhaps *A. variabilis* (Sow.) Rve. Province of Concepcion, Chili.

P. JACKSONIENSIS Lesson (Voy de la Coq. Zool. ii, p. 418, 1830). A species said to resemble *P. vulgata.* From Port Jackson, N. S. Wales.

P. GRANULOSA Lesson (Voy. de la Coq. Zool. ii, p. 422, 1830). Island of Buru. (Bourou.)

P. GIGANTEA Lesson (Voy. de la Coq. Zool. ii, p. 423, 1830). A very large shell, length 7, breadth 5 inches. It is massive, very thick, oval, convex, submedian. Muscle-impression deeply marked. Interior smooth, whitish, the cavity reddish. Outside covered with Serpula tubes. Coral reefs off Borabora, Society Is.

P. OBLONGA Perry (Conchology, t. 43; f. 4). Unknown and unknowable.

P. LAMPEDUSENSIS De Greg. An unfigured, imperfectly described form, from the Sea of Lampedusa. (Bull. Soc. Mal. Ital., x, p. 121.)

P. GRANULATA Philippi.

Shell ovate-elliptical, depressed, brown, tessellated with little-conspicuous brown spots; roughened by elevated very close, minutely granose radiating lines. Apex at the front fifth of the length. Interior bluish, margin brown, articulated with whitish. Length 10, breadth 7¾, alt. 2½ lines. (*Ph.*)

China.

P. granulata PH., Zeitschr. f. Mal. 1848, p. 162.
Compare *Acmæa schrenckii* and *concinna.*

P. ALBA Hombron & Jacquinot.

Shell ovate-elliptical, depressed, white; having sharp carinated crests, toothing the margin; apex excentral, obtuse. Interior white, shining, the depth and the margin submaculated with black. Length 37, breadth 30 mill. (*H. & J.* in Ann. des Sci. Nat. (2), xvi, 1841, p. 190.)

Tahiti.

P. TESSELLATA Hombron & Jacquinot.

Shell elevated-elliptical, convex-conic; yellow outside and inside, ornamented with undulating, transverse black-reddish lines and bands; having close radiating striæ, the chief of which number about 32. Vertex acute, gray; cavity of the same color. Length

28, breadth 22 mill. (*H. & J.* Ann. Sci. Nat. (2), xvi, p. 190, 1841.)

Island of Mangareve.

P. OBSCURA Hombron & Jacquinot.

Shell elevated-elliptical, convex, obliquely conical, blackish-brown, lightly tessellated with whitish.

Inside bluish-white, the cavity reddish-black. Vertex excentral, subobtuse. Margin entire, encircled by a black zone within. Length 26, breadth 20 mill. (*H. & J.* in Ann. des Sci. Nat. (2) xvi, p. 191, 1841.)

Talcahuano, Chili.

Evidently an *Acmæa* or a *Scurria.*

P. LUCTUOSA Hombron & Jacquinot.

Shell very angular, star-shaped, elliptical, rough, brown-black; with separated radiating ribs, of which 5 are principal and 9 or 10 adjunct, all extending beyond the margin. Vertex obtuse, excentral. Inside shining, white-bluish; margin angular, black-zoned. Length 33, width 28 mill. (*H. & J.* in Ann. Sci. Nat. (2) xvi, p. 191, 1841).

Mindanao.

May be the same as *Acmæa saccharina.*

P. CRUENTATA Hombron & Jacquinot.

Shell oval, convex-depressed, white, painted with white and brown radiating bands, ribs close, numerous, radiating; vertex brown-blackish, acute, central: margin unequally undulating, subdenticulate; inside pearly, deeply blood-stained. Length 24, breadth 19 mill. (*H. & J.*, Ann. Sci. Nat. (2), xvi, p. 191, 1841.)

New Guinea.

PATELLOIDES ANTARCTICA Hombron & Jacquinot.

Shell convex-oval, smooth, brownish-olive, painted around the margin with greenish-white bands. Vertex incumbent; inside white in front, sooty behind, the cavity rufo-castaneous; edge entire, encircled by a black zone spotted with white. Length 31, breadth 23 mill. (*H. & J.* in Ann. Sci. Nat. (2), xvi, p. 192, 1841.)

Auckland Is.

P. CALLOSA Hombron & Jacquinot.

Shell oval, depressed; radiately painted with black and white bands; inside white, the depth thickened, callous, white. Apex elevated, very acute; margin entire, colored with alternate black and white lines. Length 18, breadth 14 mill. (*H. & J.* in Ann. Sci. Nat. (2), xvi, p. 192, 1841.)

Varao.

P. ORICHALCEA Philippi.

Shell ovate-elliptical, depressed-conoid, rather thin, pellucid, pale corneous; having about 20 narrow, reddish-brown low radiating ribs, and sometimes reddish-brown decurrent streaks in the interstices. Interior beautiful silvery-golden, the center pale brown; apex at ¼ to ⅓ of the length. Margin subdentate by the ribs. Length 18, breadth 14, alt. 5½ lines. (*Ph.*)

New Zealand.

P. orichalcea PHIL., Zeitschr. f. Mal. 1848, p. 163.

Compare *P. radians* and its varieties.

P. RETICULATA Anton. Oval, widened in front; apex posterior; strong-shelled; finely longitudinally striated. Ground-color yellowish-white, with brown reticulated markings, which in front flow into a broad ray. Length 7, breadth 7½ lines. (*Anton,* Verzeichniss, p. 25, 1839.)

P. SOLIDA Anton. Oval, strongly arched; apex nearly in the middle. Strong-shelled, ribbed, irregularly rayed and flecked with white and brown. Muscle-impression very large. Length 6, breadth 4½ lines. (*Anton,* Verzeich. p. 25.)

P. OLIVACEA Anton. Refers to Gualtieri pl. 8, fig. R. No description. The cited figure is unidentifiable. (Verzeich. p. 26.)

P. LINEOLATA Anton. Oval, high. Apex at the first third of the length; white, with many small lines and strokes. Cavity of the inside brown, otherwise white within, Length 6, breadth 4 lines. (*Ant.,* Verzeich. p. 26.)

P. ALBA Anton. No description. Refers to Gualt. t. 8, fig. L, = *P. caerulea!* (Verzeich. p. 26.)

P. SERPULAEFORMIS Anton. Oval, pretty high, with 10 ribs, some projecting over the margin; brown; inside yellowish-white. Muscle-impression with brown vermiform lines. Length 10, breadth 8½ lines. (Verzeich. p. 26.)

P. PURPURASCENS Anton. Oval, with 19 equal ribs, which slightly project at the margin. Apex moderately high, blackish-red. Apex and ribs yellow, showing on the inside. Length 11, breadth 9 lines. (Verzeich. p. 26.)

P. CONICA Anton. Equals *P. vulgata* Lam., etc., the varieties with equal broad and somewhat separated ribs and conic form. Blainv. 49, 1; Mart. 1, 38. (*Anton*, Verzeich. p. 26.)

P. ALBESCENS Anton. Oval, nearly round, conical; apex wart-like, inclined forward. Very finely longitudinally, and still more finely transversely striated. Yellowish-white; inside pure white, with a brown margin around the muscle-impression, below the margin horn-colored. Length 1, breadth 11 lines. *Antilles*. (*Anton*, Verzeich. p. 26.)

P. DICHOTOMA Anton. Oval, conical, white with brown longitudinal lines, which below mostly divide gable-wise [∧-shaped]. Interior whitish. Length 6½, breadth 5¼ lines. Has the muscle-impression of Patella and the outward aspect of Siphonaria. (*Anton*, Verzeich. p. 26.)

This may possibly be *Acmæa cubensis* Rve. (*H. A. P.*)

P. (? TOREUMA, VAR.) TENUILIRATA Carpenter.

Shell much depressed, oblong, diaphanous; corneous, irregularly flamed with brownish-purple; about 22 very delicate liræ, the interstices obsoletely striated; apex subprominent, situated about at the front fifth. Interior very iridescent. Length 1·38, breadth 1, alt. ·28 inch. (*Cpr.*)

Monterey (Hartweg) in Cuming Coll.

This shell appears to agree with *P. toreuma* Reeve in all essential respects; but instead of the fine regular striæ of that species, there are a few delicate principal ribs, with obsolete striæ between. As its neighbor *P. oregona* sometimes developes large ribs, and is at other times nearly smooth, this has not been considered a sufficient difference to constitute a species until more is known of its variable powers. (*Cpr.* in P. Z. S. 1855, p. 233.)

P. RUSTICA Linn. BORN has attempted the identification of this perplexing limpet, his selection being perhaps the *P. lusitanica*. His description is on p. 426 of the Mus. Cæs., pl. 18, fig. 11. See under *P. lusitanica* and *P. neglecta*, this volume.

P. LACINOSA Linné., Syst. x, p. 781. Unidentified ; *? possibly=P. stellæformis.*

P. TUBERCULATA Linné., Syst. x, p. 782. Unidentifiable.

All Linnæan *Patellæ* other than these two and those referred to in the foregoing text of this volume, belong to other groups, such as *Fissurellidæ, Calyptræidæ, Capulidæ, Siphonariidæ, Ancylidæ, (q. v.)*

P. ISLANDICA Gmel., Syst. Nat. xiii, p. 3698, ?=*P. vulgata.*

P. RUBRA Gmel., p. 3700. Unidentified.

P. HEPATICA Gmel., p. 3700. Unidentified.

P. FUSCESCENS Gmel., p. 3701. Unidentified.

P. MACULOSA Gmel., p. 3701. Unidentified.

P. ROTUNDATA Gmel., p. 3701. Unidentified.

P. CORRUGATA Gmel., p. 3702. Unidentified.

P. ALBORADIATA Gmel., p. 3702. Unidentified.

P. OLIVACEA Gmel., p. 3702. Unidentified.

P. CEREA Gmel., p. 3702. Unidentified.

P. IMPRESSA Gmel., p. 3702. Unidentified.

P. AURANTIA Gmel., p. 3703. Unidentified.

P. MELANOZONIAS Gmel., p. 3703. Unidentified.

P. OCULATA Gmel., p. 3703. Unidentified.

P. OCHROLEUCA Gmel., p. 3703. Unidentified.

P. DENTICULATA Gmel., p. 3703. Unidentified.

P. NODULOSA Gmel., p. 3703. Unidentified.

P. CINEREA Gmel., p. 3704. Unidentified.

P. EXALBIDA Gmel., p. 3704. Unidentified.

P. LÆVIS Gmel., p. 3704. Unidentified.

P. ARGENTEA Gmel., p. 3704. Unidentified.

P. CUPREA Gmel., p. 3704. I do not know this apparently well-marked species. It may be a form of *P. magellanica.*

P. SANGUINEA Gmel., p. 3705. Unidentified.

P. INÆQUALIS Gmel., p. 3705. Unidentified.

P. FLAVEOLA Gmel., p. 3705. Unidentified.

P. INFUNDIBULUM Gmel., p. 3705. Unidentified.

P. CYATHUS Gmel., p. 3705. Unidentified.

P. ULYSSIPONENSIS Gmel., p. 3706. This is a form of *P. cœrulea*.

P. MELANOGRAMMA Gmel., p. 3706. Unidentified.

P. REPANDA Gmel., p. 3707. Probably *P. aspera* of the Mediterrean.

P. ANGULOSA Gmel., p. 3707. Unidentified.

P. TIGRINA Gmel., p. 3707. Unidentified.

P. MONOPIS Gmel., p. 3707. ?= *P. oculus* Born.

P. CHLOROSTICTA Gmel., p. 3707. ?= *P. cœrulea var. crenata*.

P. MARGARITACEA Gmel., p. 3707. ?= *P. cœrulea*.

P. TENUIS Gmel., p. 3708. ?= *P. cœrulea*.

P. PLICARIA Gmel., p. 3708. ? *P. barbara* Linn., p. 96.

P. STANNEA Gmel., p. 3709. ?= *P. ænea* Martyn.

P. FASCIATA Gmel., p. 3713. Unidentified.

P. ELEGANS Gmel., p. 3713. Unidentified.

P. SQUAMOSA Gmel., p. 3713. Unidentified.

P. SQUALIDA Gmel., p. 3714. Unidentified.

P. CROCEA Gmel., p. 3714. Unidentified.

P. CANDIDA Gmel., p. 3714. Unidentified.

P. MINIMA Gmel., p. 3714. ? *Acmæa virginea* Müll.

P. TRANQUEBARICA Gmel., p. 3714. An unidentifiable Oriental *Acmæa*.

P. SURINAMENSIS Gmel., p. 3716. Unidentified.

P. VITELLINA Gmel., p. 3716. Unidentified.

P. LÆVIGATA Gmel., p. 3717. Unidentified.

P. CITRINA Gmel., p. 3720. Unidentified

P. GUTTATA Gmel., p. 3721. Unidentified.

P. SCUTIFORMIS Gmel., p. 3721. Unidentified.

P. CRATICULATA Gmel., p. 3722. Unidentified.

P. CRUENTATA Gmel., p. 3722. Unidentified.

P. PAPYRACEA Gmel., p. 3722. Unidentified.

P. CYLINDRICA Gmel., p. 3722. Unidentified.

P. DECUSSATA Gmel., p. 3723. Unidentified.

P. HÆMOSTICTA Gmel., p. 3723. Unidentified.

P. ASTEROIDES Gmel., p. 3723. Unidentified.

P. RUBELLA Gmel., p. 3723. Unidentified.

11

P. SPECTABILIS Gmel., p. 3723. Unidentified.

P. CONSPURCATA Gmel., p. 3724. Unidentified.

P. ATRA Gmel., p. 3724. Unidentified.

P. SPECULARIS Gmel., p. 3724. Unidentified.

P. CANESCENS Gmel., p. 3724. Unidentified.

P. VIRESCENS Gmel., p. 3724. Unidentified.

P. PULLA Gmel., p. 3725. Unidentified.

P. REVOLUTA Gmel., p. 3725. Unidentified.

P. SQUAMATA Gmel., p. 3725. Unidentified.

P. TESTACEA Gmel., p. 3725. Unidentified.

P. CAPILLARIS Gmel., p. 3725. Unidentified.

P. GLAUCA Gmel., p. 3725. Unidentified.

P. OBSCURA Gmel., p. 3726. Unidentified.

P. EXOLETA Gmel., p. 3726. Unidentified.

P. AFFINIS Gmel., p. 3726. Unidentified.

P. FUSCATA Gmel., p. 3726. Unidentified.

P. MELLEA Gmel., p. 3726. Unidentified.

P. GUINEENSIS Gmel., p. 3726. Unidentified.

P. COMPLANATA Gmel., p. 3726. Unidentified.

P. NAVICULA Gmel., p. 3727. ?= *P. miniata.*

P. CINGULATA Gmel., p. 3727. Unidentified.

A portion of these species of Gmelin may belong to *Siphonariidæ,* although all species belonging elsewhere than in *Patellidæ* and *Acmæidæ* have been purposely omitted in the above list.

P. LUTEOLA Lam., An. s. Vert. vi, p. 327. Unidentified.

P. TUBERCULIFERA Lam., An. s. Vert. vi, p. 333. Unidentified.

P. RADIATA Born, Test. Mus. Cæs. Vindob., t. 18, f. 10. (*P. virgata* Gmel., Syst. xiii, p. 3727). Unidentified.

P. BORNIANA Helbling, Abhandl., p. 106, t. 1, f. 7.=*Acmæa testudinalis.*

P. ZONATA Schubert & Wagner, Conchyl. Cab., p. 125, t. 229, f. 4056, 4057. Unidentified.

P. VIRGATA Donovan, in Rees' Encycl., Conch., pl. xvi. Unidentified.

P. AURICULA Donovan, in Rees' Encycl., Conch. pl. xvi. Unidentified.

P. STRIGATA Donovan, in Rees' Encycl., Conch. pl. xvi. Unidentified.

P. ONYCHITES Menke, Moll. Nov. Holl. Spec., p. 34. Unidentified. Western Australia.

P. PULCHELLA Blainville, Dictionnaire des Sciences Naturelles, xxxviii, 1825, p. 92. This may be Patina pruinosa.

P. LOBATA Blainv., L. c., p. 93. This is either a strongly costate form of Nacella mytilina, such as N. compressa of Rochebr. & Mabille, or a nearly smooth N. deaurata.

P. CASTANEA Blainv. L. c., p. 94. Unidentified.

P. VIRIDESCENS Blainv., L. c., p. 95. Probably a form of Nacella ænea. Falkland Is.

P. TENUISTRIATA Blainv., L. c., p. 96. Unidentified.

P. NIGRA Blainv., L. c., p. 96. Unidentified.

P. MACULATA Blainv., L. c., p. 97. Probably it is P. capensis or P. variabilis. From the Cape.

P. DEPRESSA Blainv., L. c., p. 97. Unidentified.

P. LUGUBRIS Blainv., L. c., p. 99. A Helcioniscus, like H. amussitatus. Rv. From the Moluccas.

P. AURANTIACA Blainv., L. c., p. 99. Unidentified. Habitat unknown.

P. VARIEGATA Blainv., L. c., p. 100. Unidentified. From Botany Bay.

P. SQUAMA (Gualt., t. 8, f. L.) Blainv., L. c., p. 101.==P. cærulea.

P. GRISEA Blainv., L. c., p. 102.==P. aspera. From Greece.

P. ALBORADIATA Blainv., L. c., p. 102. Unidentified.

P. PARALLELOGRAMMICA Blainv., L. c., p. 103. Unidentified.

P. HEPTAGONA Blainv., L. c., p. 104.==Acmæa saccharina.

P. CHILENSIS Blainv., L. c., p. 104.==Siphonaria?

P. CONICA Blainv., L. c., p. 107. Island of Maria. This is a large species, perhaps Helcioniscus argentatus.

P. CAMPANIFORMIS Blainv., L. c., p. 108.==Siphonaria?

P. CARDITOIDEA Blainv., L. c., p. 110.==Nacella deaurata?

P. RARICOSTA Blainv., L. c., p. 110. Unidentified.

P. SOLIDA Blainv., L. c., p. 110. Unidentified. Australia.

P. RUBROAURANTIACA Blainv., L. c., p. 111. Unidentified.

P. LATICOSTATA Blainv., *L. c.*, p. 111. Probably this is *P. neglecta* Gray.

P. NIGROALBA Blainv., *L. c.*, p. 111. Unidentified. Cape Good Hope.

P. PERONII Blainv., *L. c.*, p. 111. Unidentified. Port of King George, Australia.

P. MADAGASCARENSIS Blainv., *L. c.*, p. 112. Unidentified.

P. VIOLACEA Blainv., *L. c.*, p. 112. Perhaps—*P. vulgata*.

P. ATROMARGINATA Blainv., *L. c.*, p. 113.= *P. granularis*.

P. SUBGRANULARIS Blainv., *L. c.*, p. 113.=*P. lusitanica*.

P. ZEBRA Blainv., *L. c.*, p. 115. Unidentified.

P. UNGULIFORMIS Blainv., *L. c.*, p. 115. Unidentified.

Family *TITISCANIIDÆ.*

Animal limaciform, naked, without a shell; radula rhipidoglosate, neritoid, but lacking median teeth like the *Neritopsidæ.*

This shell-less, limaciform type belongs, as its internal anatomy and the armature of the radula show, to the group of the *Neritacea;* and the lack of median teeth places it evidently in the *Neritopsoid* division of that group.

Genus TITISCANIA Bergh, 1890.

Titiscania BERGH, Morphol. Jahrbuch, xvi, (1), p. 3. Type, *T. limacina.*

T. LIMACINA Bergh. Plate 41; pl. 39, figs. 28, 29.

Form quite *limax*-like, long-oval, convex above, flat beneath, rounded in front and behind. The color above is clear yellowish, whiter in the middle, where the violet-gray entrails show through, and with a few whitish transverse bands. The back is quite even, only with a few white salient points; anteriorly above is the transverse branchial slit, the end of the gill projecting out of the slit in several individuals. The gill has 35 to 40 laminæ. In front of the gill-slit is the flat and rather wide head with rather long, pointed tentacles, having the black, nearly sessile eyes at their outer bases. In the male there is behind the right eye a little whitish hump or pit (opening) for the end of the seminal duct. The entire length is 10½ mill.

Camiguin, Philippines; Mauritius.

Titiscania limacina BERGH in Semper's Reis. Archip. Phil. II, ii, Heft ix, t. xli, f. 10; Morphol. Jahrb. xvi, p. 3. t. 1, 2.

The coloration of the specimens from the Philippines is described above. The specimens from Mauritius were mostly smaller (length 9, breadth scarcely 4, alt. 2–2½, length of tentacle 1½ mill.) the color whitish, the entrails not showing through, with the exception of the anterior female genital mass which conspicuously shows through.

Explanation of figures, plate 41.

Fig. 53. Anterior end of animal from below, showing front gill-commissure with *retracted* gill, then tentacles, oral aperture and sole of foot.

Fig. 54. Anal papilla.

Fig. 55. Alimentary canal.

Fig. 56. The living animal, enlarged after a drawing by Semper.

Fig. 57. Tentacle with eye.

Fig. 58. Radula (x100).

Fig. 59. *a*, Lateral tooth lying prostrate; *b*. Three inner teeth of the second row; *c*. Two inner supporting plates.

Fig. 60. Outer plates.

Fig. 61. Innermost of the outer plates.

Fig. 63. Outermost plate.

Fig. 64. Outer end of a row of teeth.

Fig. 65. A few teeth from the same.

Plate 39, figs. 28, 29. Nerve ganglia.

Appendix.

Family ACMÆIDÆ Cpr.

Dr. THIELE, in his continuation of Troschel's *Das Gebiss der Schnecken*, recognizes two subfamilies of *Acmæidæ*, identical with those adopted by me on p. 6 of this volume.

ACMÆA VIRIDULA Lam. (p. 32). Reeve, on a suppressed page of the Conchologia Iconica, names this species *P. nivalis.*

ACMÆA VARIABILIS Sowb. (page 34.)

Philippi changed the name of his *P. lineata* to *P. grammia* in the index to the *Abbildungen*. This name will also fall into line in the procession of synonyms after *A. variabilis.*

ACMÆA ANTILLARUM Sowb. (p. 38).

This name will take precedence over that of *candeana, elegans,* etc., the former of which was used in the text, p. 38.

Sowerby's figure of *Lottia antillarum,* on the plate first cited below, is an excellent and characteristic picture of this species in its finest development.

The synonymy will stand as follows:

Lottia antillarum SOWERBY, Genera of Shells, fig. 4. (Issued before 1831.)

Lottia antillarum SOWERBY, A Concholog. Manual, p. 59, fig. 231, 1839. (A somewhat different color-form.)

Lottia antillarum Sowb., REEVE, Conchol. System., pl. cxxxvii, f. 4 (printed from same plate as Sowerby's *Genera*) 1842.

Acmæa antillarum Sowb., PILSBRY, *The Nautilus,* Dec., 1891, p. 85.

Patella tenera C. B. ADAMS, Proc. Bost. Soc. N. H. ii, p. 8 (1845).

Patella tenera Ad., REEVE, Conch. Icon. fig. 104.

Patella candeana ORB., Moll. Cuba, ii, p. 199, atlas pl. 25, figs. 1–3.

Acmæa candeana Orb., DALL, Catal. Mar. Moll. S. E. U. S., p. 159.

Acmæa candeana Orb., PILSBRY, Manual of Conchology, xiii, p. 38, pl. 5, figs. 91–95, and pl. 42, figs. 92–95.

? *Patella (Acmæa?) elegans* PHILIPPI, Abbild. u. Beschreib. iii, p. 34, *Patella* p. 6, pl. 2, fig 2 (1846).

? Not *P. antillarum* Sowb., PHILIPPI, Abbild. iii, *Patella,* pl. 2, fig. 12.

ACMÆA ONYCHINA Gould. Pl. 73, figs. 96, 97, 98.

The original figures of this species are not very satisfactory, although Gould's description is very good. Figures are here given drawn from specimens lately collected at Bahia, Brazil.

The diagnostic characters are, besides the orbicular and rather depressed form, the sculpture, which consists of a variable number of low, rounded radiating ribs (often nearly obsolete), the *entire surface being finely radially striated*. The erosion of the surface is generally, so far as my specimens show, extensive, the sculpture being lost to a corresponding degree. The interior is characterized by a *black-brown spatula* which often has a "tail-piece" similar to the usual head-segment, and nearly always there are *broad rays extending to the front and posterior margins from the central spatula*. The marginal border is either black or tessellated. This species has a wide distribution along the east coast of South America, extending as far south as Santa Caterina, in S. Lat. 32° 30′, where it has been collected by Dr. H. von Ihering (see Dall, in *The Nautilus*, Aug., 1891, p. 44). See also the locality of the synonymous *P. mülleri* Dkr., this volume, page 43. The name *Acmæa subrugosa* was published by Orbigny prior to the Gouldian name *onychina*, but the diagnosis given is scarcely sufficient for recognition. It is as follows :

A. *testa ovato-convexiuscula, striato-costata; striis inæqualibus; albido-virescente zonis fuscis radiata; intus fusco; margine subcrenato, lutescente, fusco-maculato.* Diam. 18 mill., alt. 8 mill. (Orb.)

Mr. E. A. Smith has recently described a form closely allied to this species, from the island Fernando Noronho. It should probably be considered a variety of the *A. onychina*.

Var. NORONHENSIS E. A. Smith. Pl. 73, figs. 93, 94.

Shell ovate, wider behind, moderately elevated, blackish, painted with pale rays, eroded and black at the apex, which is situated a little in front of the middle; radiately delicately striated, sculptured with lines of increment.

Interior blackish within the muscle-scar, toward the apex having a thin whitish callus; outside of the muscle-scar, nearly to the edge, it is bluish-white; at the margin narrowly edged with black; having a wide dark ray extending from apex to margin in front, and a broader one behind. (*Smith.*)

Length 24, width 19, alt. 9 mill.

Island of Fernando Noronho, off Brazil.

Acmæa noronhensis Smith, Journ. Linn. Soc. Lond. xx, p. 495, t. 30, f. 3, 3a (1891.)

This species has a smoother surface than *A. subrugosa*, d'Orbigny (= *Lottia onychina* Gould), from Rio Janeiro.   Like that species, however, it has in the interior a broad obscure ray from the apex to the margin in front and a broader one at the opposite end.   These rays, however are more distinct in the present species than in the Brazilian shell.   The external radiating striæ being very fine, do not, as a rule, produce a crenulated margin, but in some instances a slight crenulation occurs.   The surface within the muscle scar is almost black, forming a marked contrast to the pallid space between it and the black margin, shells found attached to rocks, when placed upon a flat surface, rest upon the anterior and posterior margins only, so that the sides are slightly raised. (*Smith.*)

SCURRIA ZEBRINA Lesson. (page 63.)

Following Dall, I included " *Patella concepcionensis* Lesson " in the synonymy of *zebrina* on page 63. Lesson described no such species, his name being "*concepsionis*" ; and its pertinence to *zebrina* is doubtful. See at foot of page 155, this volume.

New Zealand Acmæidæ.

ACMÆA LACUNOSA Reeve (p. 52).

Hutton omits this name from his latest revision of the New Zealandic Acmæids (Proc. Linn. Soc. N. S. Wales, ix, p. 372, 1884). He considers *A. corticata* a good species. Figures 9, 10, 11 of my plate 37 represent specimens of *corticata* received from Hutton, and figs. 7, 8 of pl. 37 are Reeve's originals of *lacunosa*.

A. CINGULATA Hutton (p. 53).

Add to references : Hutton, N. Z. Journ. of Sci. i, p. 477, 1883 ; Proc. Linn. Soc. N. S. Wales, ix, p. 372.

A. RUBIGINOSA Hutton (p. 53).

This unfigured species I have not seen. It may = *A. lacunosa*. Hutton considers *Patella campbelli* Filhol (Compt. Rend. xci, p.

1095, 1880) a probable synonym. *P. campbelli* is insufficiently diagnosed.

A. CONOIDEA Q. & G. (p. 53).

Hutton records this from Banks' Peninsula, N. Z.

A. FLAMMEA Q. & G. (p. 57).

Hutton reports this from New Zealand, Auckland to Dunedin.

A. CRUCIATA Linné. Pl. 73, fig. 95.

Shell oval, moderately solid, varying in elevation from subdepressed to subconical, usually found smooth, but rayed in fresh and perfect individuals with very fine raised striæ. Coloring rather variable, yet almost always exhibiting a more or less cruciform arrangement; when most characteristic, displaying four broad white rays upon a white spreckled ground of blackish-brown that are usually bisected, as they spread, by a short dark streak which at times becomes so broad as to produce the appearance of there being eight narrow white rays, or of a cross with white edges and a brown center; occasionally, too, there are narrower interstitial rays besides. Apex blunt, yet prominent, always white both within and without, placed at rather more than one-third the space from the narrower end. Interior with a faint central brown spatula-shaped stain, intersected by the external rays, which appear more or less visibly through the very thin white glaze that lines the rest of the cavity. Length three-fourths of an inch; breadth half an inch. Worn individuals exhibit a brown cross upon a white ground. (*Hanley.*)

Habitat unknown.

Patella cruciata L., Syst. Nat. x, p. 784.—SCHRŒTER, Einleitung in die Conchylien-Kenntniss, ii, p. 432, t. 5, f. 6.—*Acmæa cruciata* HANLEY, Shells of Linn., p. 429; Wood's Index Testac., p. 189, t. 38, f. 78, (edit. Hanley).—? *Patella insignis* MKE., Moll. Nov. Holl. Spec., p. 34; Zeitschr. f. Mal. 1844, p. 62.

Compare *Acmæa crucis* Tenison-Woods. The figure is reduced in size.

P. INSIGNIS Menke.

Shell ovate, convex, shining, subpellucid, obsoletely concentrically striated, vertically subsulcate behind, whitish, painted with five or six rather wide reticulated brown rays. Vertex excentric, margin entire, length 7·4, breadth 5·7, alt. 3 lines, (*Mke.* in Moll. Nov. Holl. Spec.. p. 34.)

Western Australia.

This is not *P. insignis* Dkr., a synonym of *P. testudinaria.*
It is probably the same as *Acmœa cruciata* Linn.

ACMÆA ARANEOSA Gould.   Pl. 73, figs. 90, 91, 92.

Shell small, thin, rounded, slightly elevated, smooth, and without ribs or sculpture; apex nearly central, obtuse.   Color pale yellow-ish-green, reticulated with very fine rusty-brown lines, branching off like rootlets towards the margin.   The interior is whitish, with a rusty ring just within the muscular impression; the edge is sharp and simple, and has a well-defined limbus, so thin as distinctly to repeat the external linear markings.

Length five-eighths of an inch; breadth half an inch; height one-fourth of an inch.   (*Gld.*)

A pretty little shell, most probably a Lotti, resembling some of the species figured by Quoy, especially his *orbicularis.*   Indeed it is so like the reticulated variety figured in Quoy's pl. 71, fig. 33, that the examination of a more extensive series might show them to be identical.   (*Gld.*)

Sooloo Sea.

Patella (*Lottia ?*) *araneosa* GLD., Proc. Bost. Soc. N. H. ii, p. 152 (1846); U. S. Exped. p. 347, figs. 450.

———   ——   —   ——

PATELLA BARBARA Linné.   (Page 96.)

No locality is given in the text for this species.   I am informed by MR. GEO. W. TAYLOR that it is abundant at the *Cape of Good Hope.*

HELCIONISCUS REYNAUDI Desh.   (Page 130).   By a typographical error this name is spelled incorrectly in the text.

Family PATELLIDÆ.

Dr. Thiele, in his continuation of Troschel's *Das Gebiss der Schnecken* (received at Philadelphia since the publication of part 50 of the MANUAL, in which my own classification of *Patellidæ* was out-lined), divides the family according to the characters of the radula into groups as seen below.   The great merit of Dr. Thiele's work consists in his demonstration of the existence of a rhachidian tooth in all groups of *Patellidæ;* in his recognition of the fact that the *primary* division of the family is into two groups, based upon the number of anterior side teeth—a generalization which I had made before seeing his publication, as will be seen by my synopsis on page 79 of this volume.

I am satisfied that no *generic* distinctions can be based upon the degree of development of the rhachidian tooth. It varies in different species from a mere rudiment to a tooth as well-developed as the laterals. It is to be expected that the examination of more material will bridge such gaps as still exist in the range of its variation.

Dr. Thiele's arrangement is as follows. I have quoted in brackets the species investigated by him of each group.

Family *Patellidæ.*

Subfamily PATELLINÆ [equals in limits and contents, division *A* of my synopsis on p. 79.]
  Genus *Ancistromesus* Dall. [P. chitonoides, P. pica.]
  Genus *Patellidea* Thiele. [P. granularis.]
  Genus *Patellona* Thiele. [P. granatina, P. plumbea?, P. adansoni.]
  Genus *Olana* Ads. [P. cochlear.]
  Genus *Cymbula* Ads. [P. compressa.]
  Genus *Patellastra* Monts. [P. lusitanica, P. guttata, P. ferruginea.]
  Genus *Patella* L. [P. tarentina, P. crenata, P. cœrulea, P. scutellaris, P. aspera, P. lugubris, P. moreleti, P. vulgata.]
  Genus *Patellopsis* Thiele. [P. ? shell unknown.]
  Genus *Helcion* Montf. [P. pectunculus Gm.=P. pectinatus Born.]
  Genus *Patinastra* Thiele. [P. pruinosa.]
  Genus *Patina* Leach. [P. pellucida, P. tella.]
Subfamily NACELLINÆ [Equals in limits and contents, division *B* of my synopsis on page 79.]
  Genus *Nacella* Schum. [P. vitrea, P. hyalina, P. mytilina.]
  Genus *Patinella* Dall. [P. deaurata, P. venosa, P. fuegiensis, P. atramentosa.]
  Genus *Helcioniscus* Dall. [P. toreuma, P. amussitata, P. testudinaria, P. exarata, P. rota.]

Different authors entertain such diverse views upon the amount of divergence sufficient to give *generic* rank to a group, that any discussion of this matter would be futile. Such differences of opinion are inevitable, and it must be left to the zoologists of the future to sift these things down to a just, convenient and uniform usage. It should be noted, however, that most of the above " genera " are founded upon the *degrees of development* of a *single* organ.

I have copied on pl. 52, fig. 6, Thiele's figure of the radula of *Patellidea granularis*. On pl. 74, fig. 1, dentition of *Patellona granatina*. On pl. 52, fig. 7, the dentition of *Olana cochlear*. On pl. 74, fig. 2, that of *Cymbula compressa*. On pl. 52, fig. 8, that of *Patellastra lusitanica*. On pl. 52, fig. 5, that of *Patella aspera*. On pl. 52, fig. 9, that of *Patellopsis* sp., a Cape species of which the shell is unknown.

Pl. 52, fig. 4, represents the dentition of *Helcion pectinatus*. Pl. 52, fig. 3, that of *Patinastra pruinosa*. Pl. 52, fig. 2, the dentition of *Patina pellucida*.

On pl. 74, fig. 3, the dentition of *Nacella vitrea* is figured. Pl. 74, fig. 4, represent that of *N. mytilina*. Pl. 74, fig. 5, that of *Patinella venosa*. Pl. 74, figs. 7, 8, that of *P. faegiensis*.

Pl. 74, fig. 6, represents the dentition of *Helcioniscus capensis*.

It is likely that naturalists of all schools could agree upon some such arrangement of this family as the following:

REVISED CLASSIFICATION OF THE PATELLIDÆ.

Family *Patellidæ*.

I. Subfamily PATELLINÆ. Lateral teeth of the radula *three* on each side, two of them anterior.
   1. Genus *Patella* L. Branchial cordon complete; apex of shell near the center.
   2. Genus *Helcion* Montf. Branchial cordon interrupted in front; apex of the shell anterior.
II. Subfamily NACELLINÆ. Developed lateral teeth but *two* on each side, of which one is anterior.
   3. Genus *Nacella* Schum. An epipodial ridge developed upon the sides of the foot; branchial cordon complete.
   4. Genus *Helcioniscus* Dall. Sides of foot smooth, with no trace of an epipodial ridge. Branchial cordon interrupted in front.

REFERENCE TO PLATES.

VOL. XIII.

PLATE 1.

Plate 12.

  12

PLATE 28.

PLATE 38.

PLATE 39.

PLATE 40.

PLATE 41.

(See also pl. 39, figs. 28, 29.)

PLATE 42.

PLATE 61.

PLATE 62.

PLATE 63.

PLATE 64.

PLATE 65.

PLATE 66.

PLATE 67.

PLATE 68.

INDEX TO VOL. XIII.

—

NOTE.—The names of valid species and varieties are printed in Roman type; of genera and other groups in SMALL CAPITALS; of synonyms in *Italic*.

190   INDEX.

NOTE.—The Parts of Vol. XIII of the MANUAL were issued to subscribers upon the following dates:

Part 49, including pp. 1–64, plates 1–15, August 3, 1891.

Part 50, including pp. 65–112, plates 16–35, November 3, 1891.

Part 51, including pp. 113–160, plates 36–55, January 30, 1892.

Part 52, including pp. 161–196, plates 56–74, March, 1892.

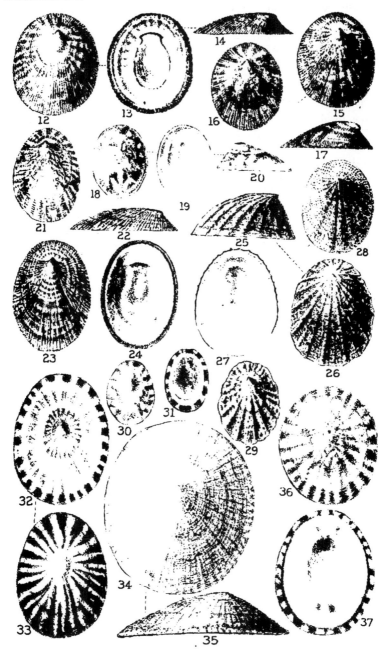

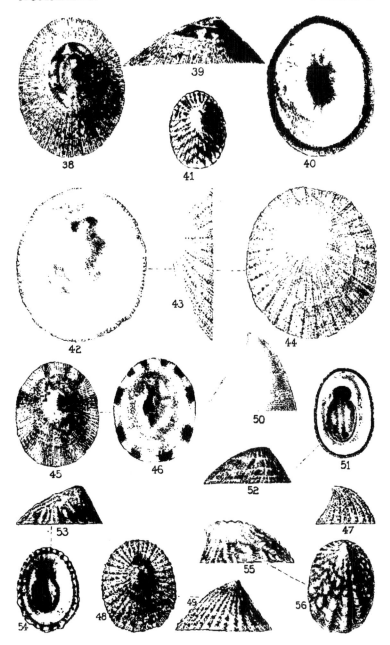

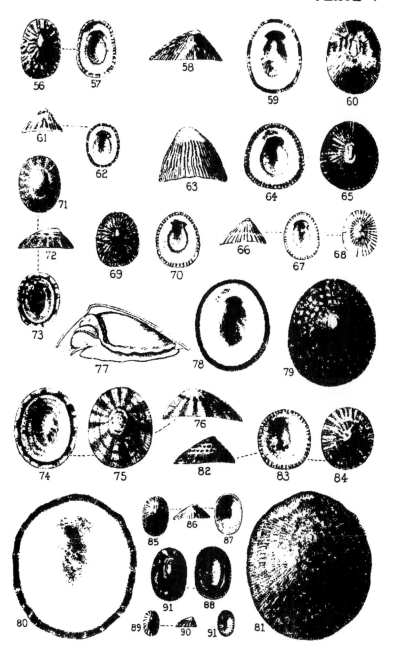

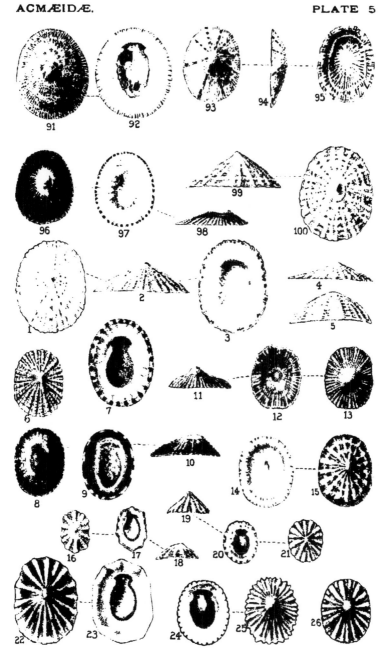

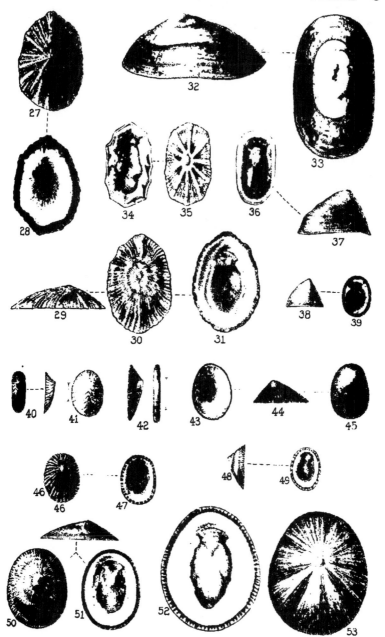

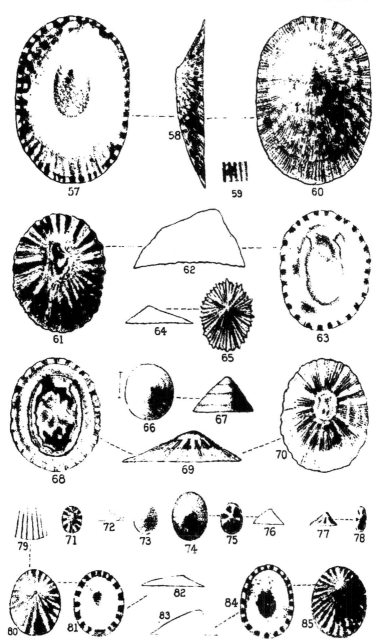

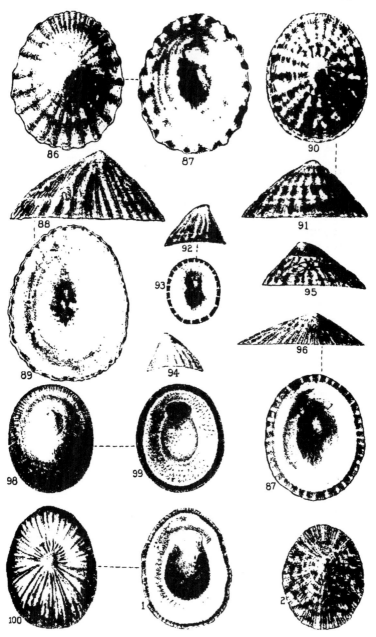

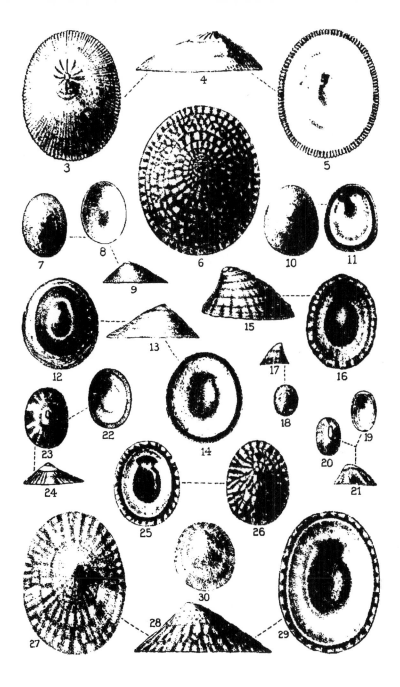

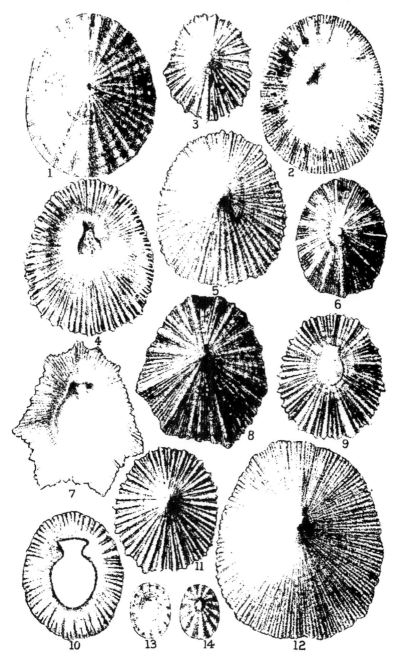

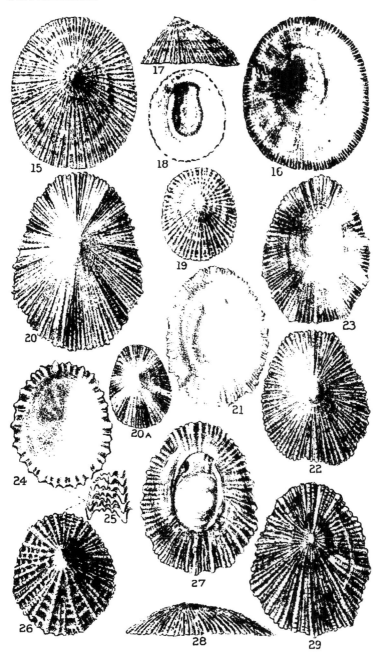

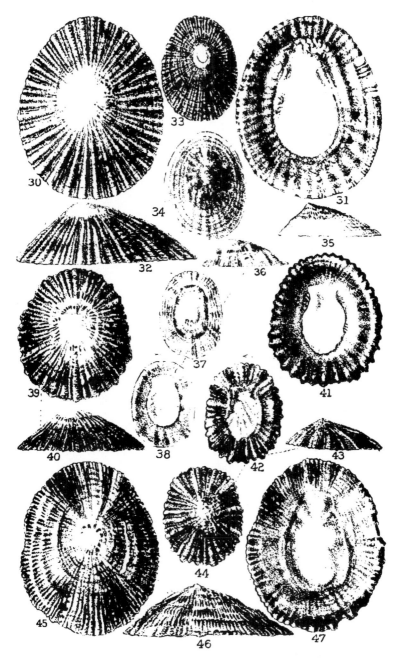

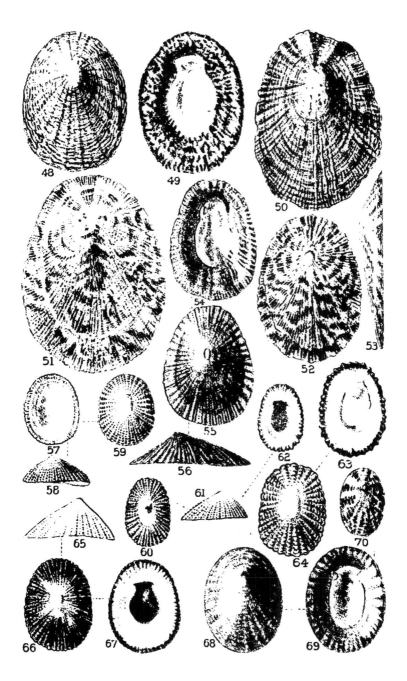

PLATE 14

PATELLIDÆ.

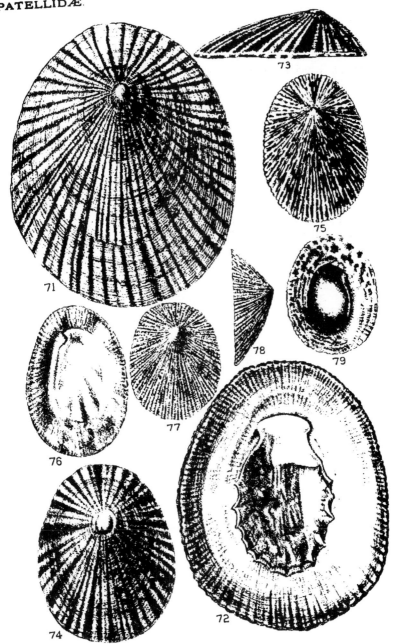

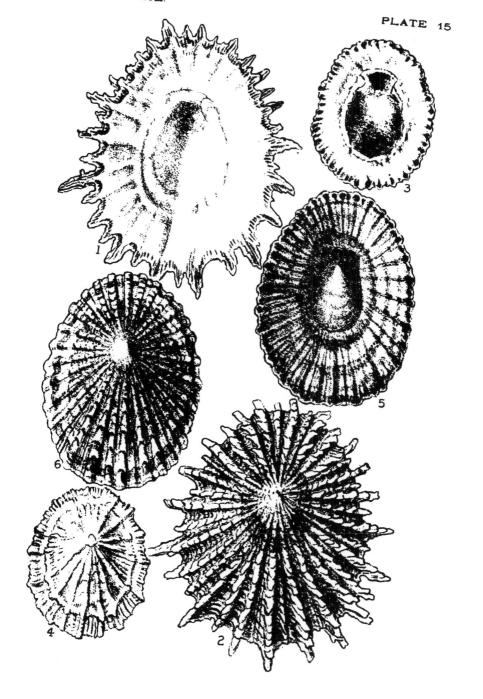

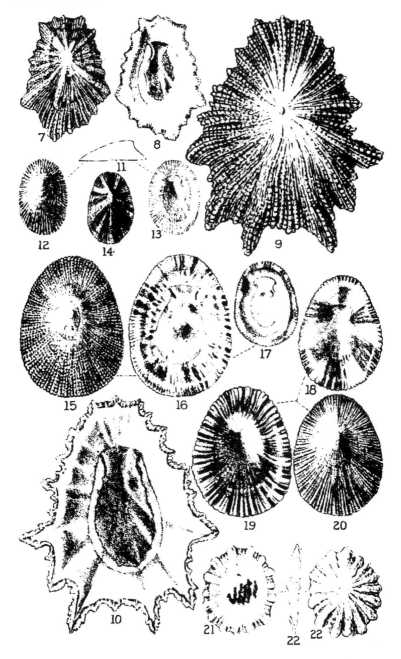

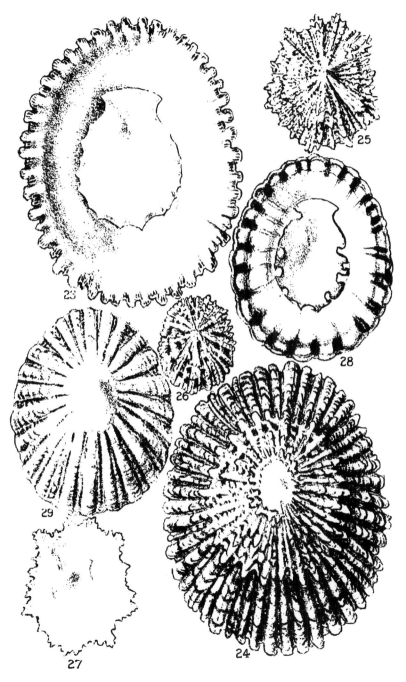

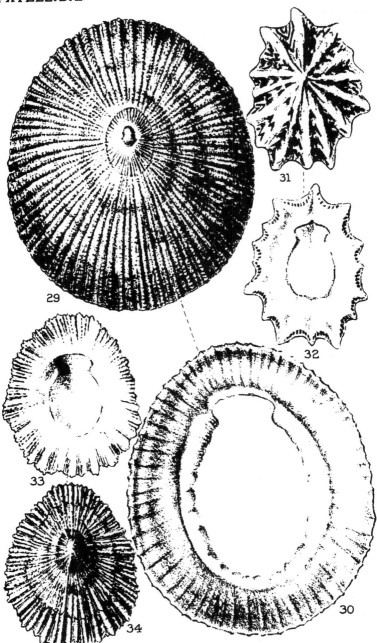

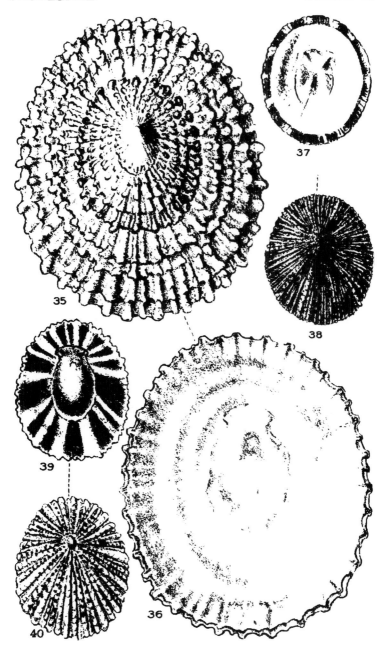

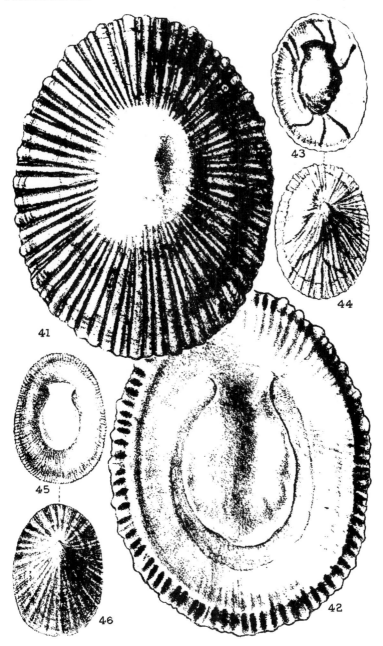

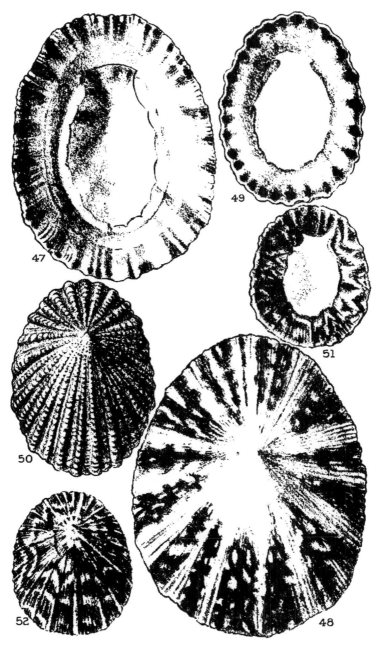

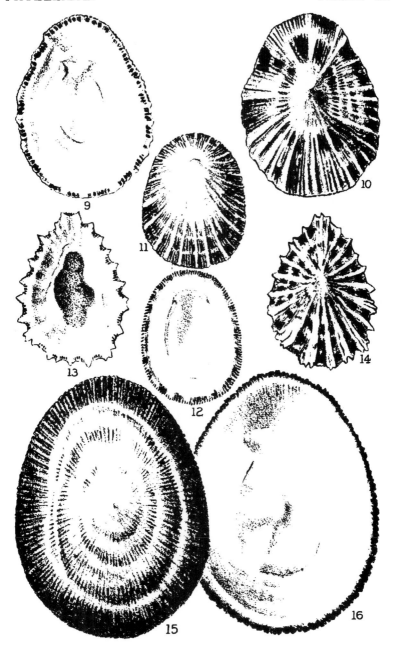

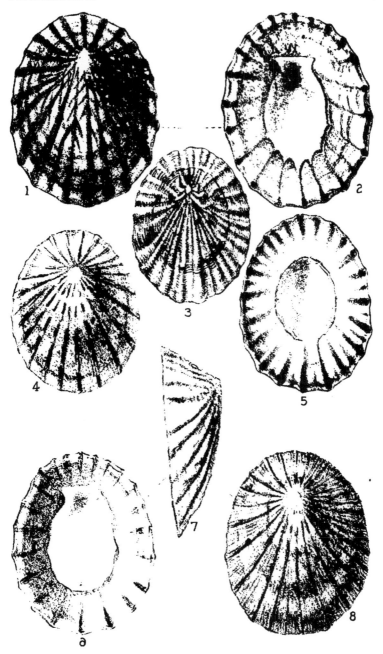

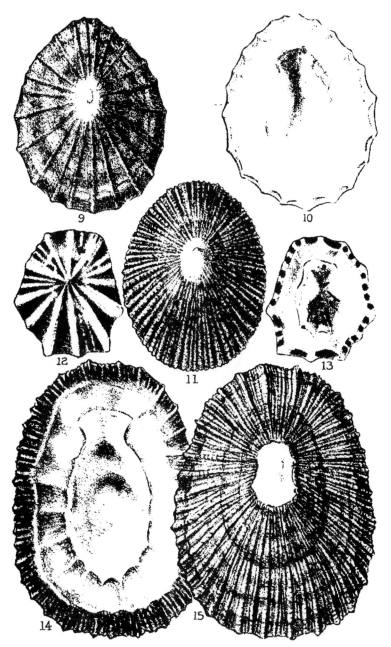

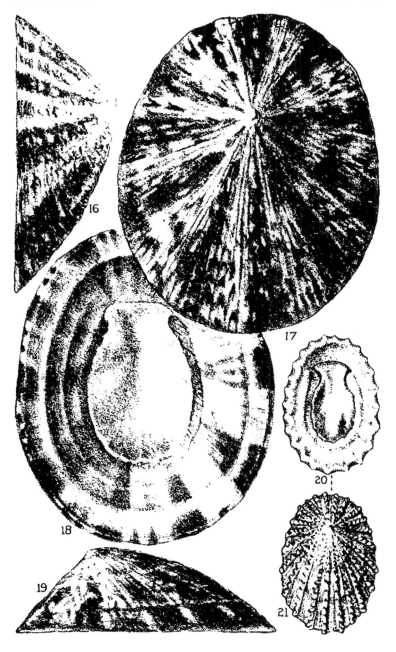

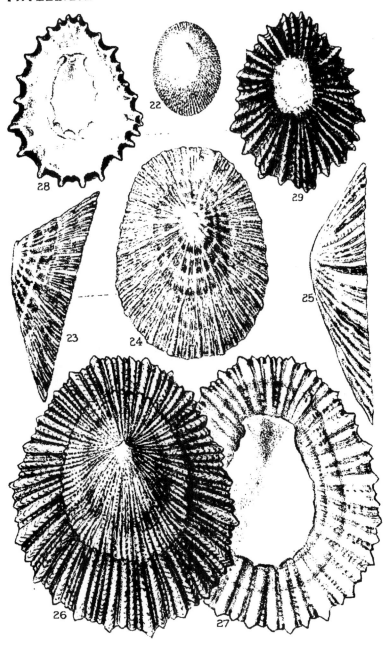

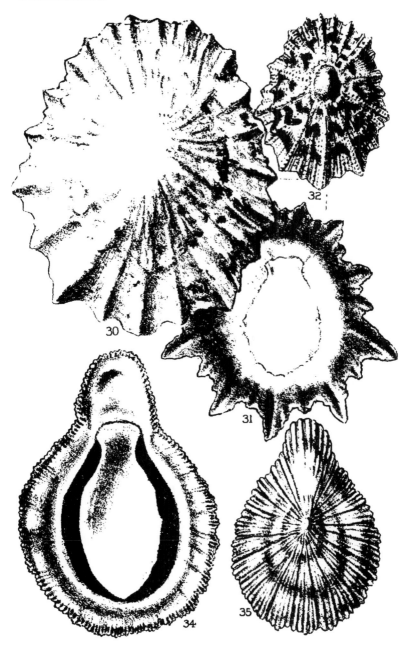

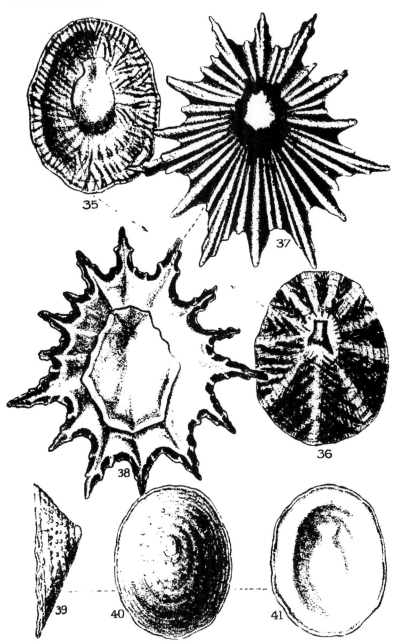

35

37

38

36

39 40 41

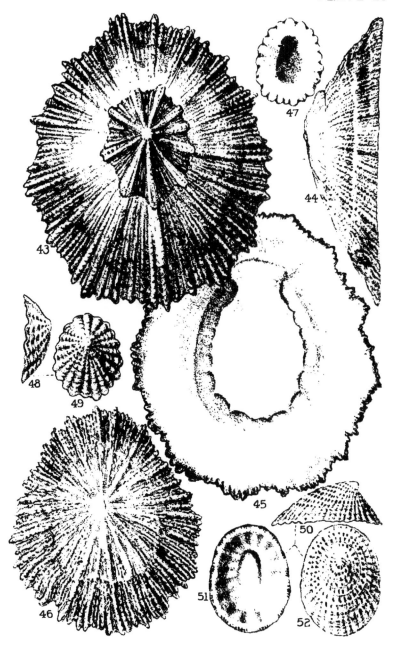

PLATE 30

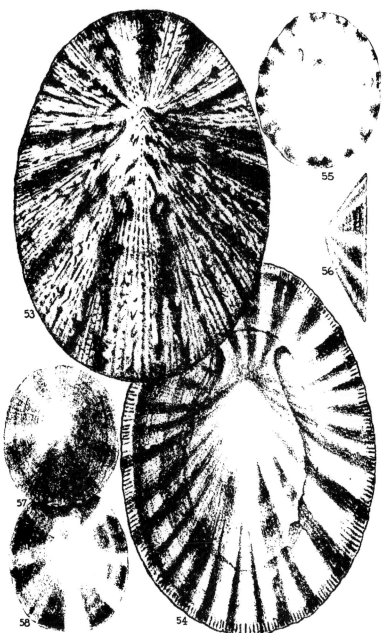

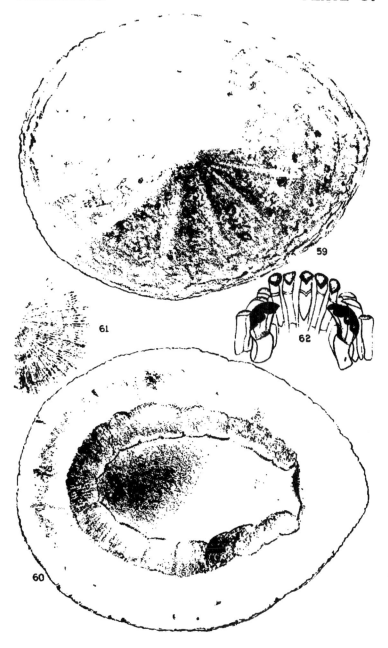

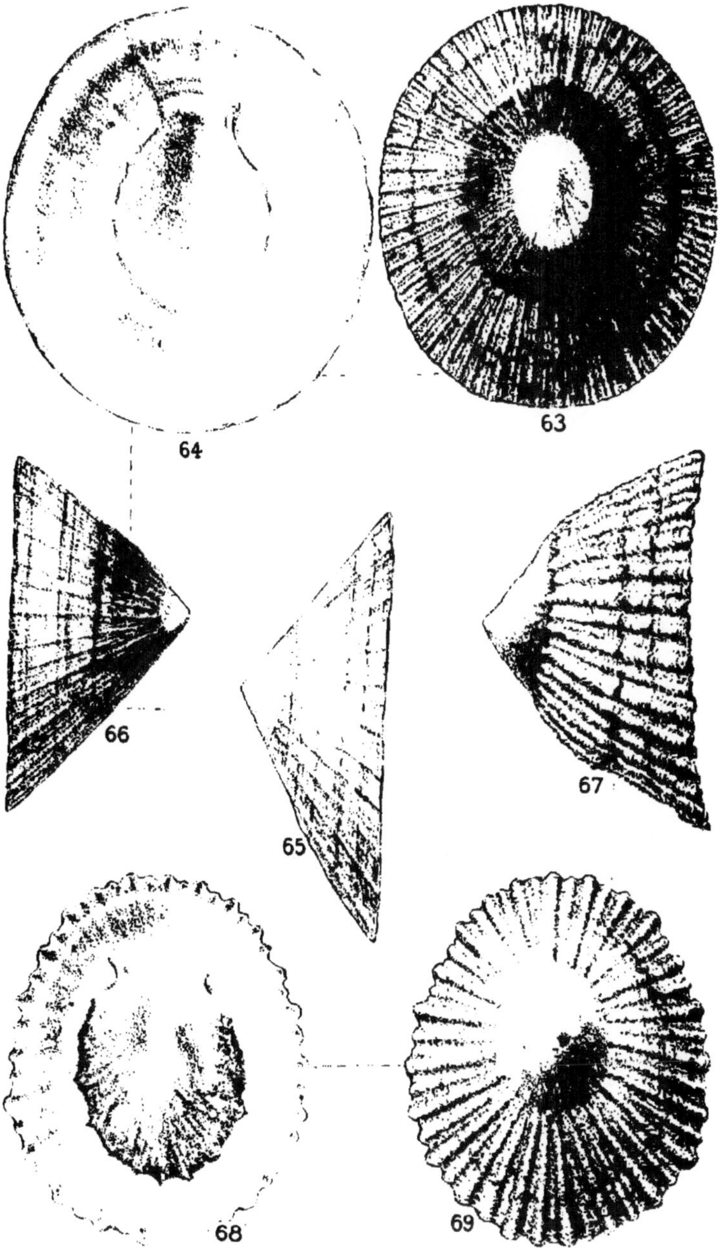

64

63

66

65

67

68

69

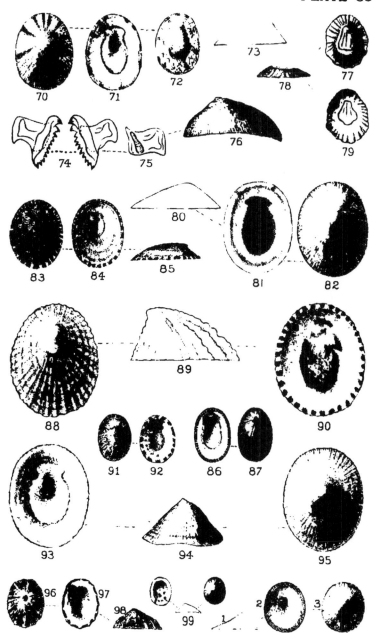

PLATE 34

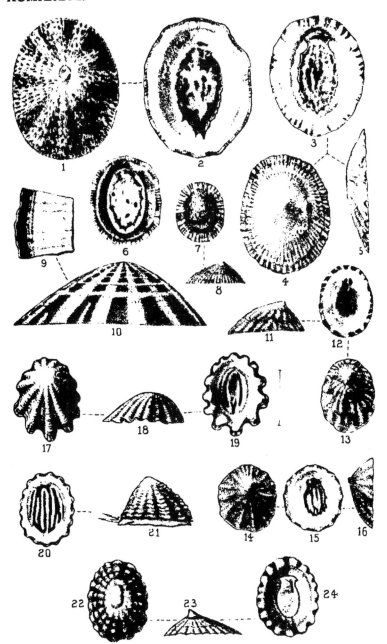

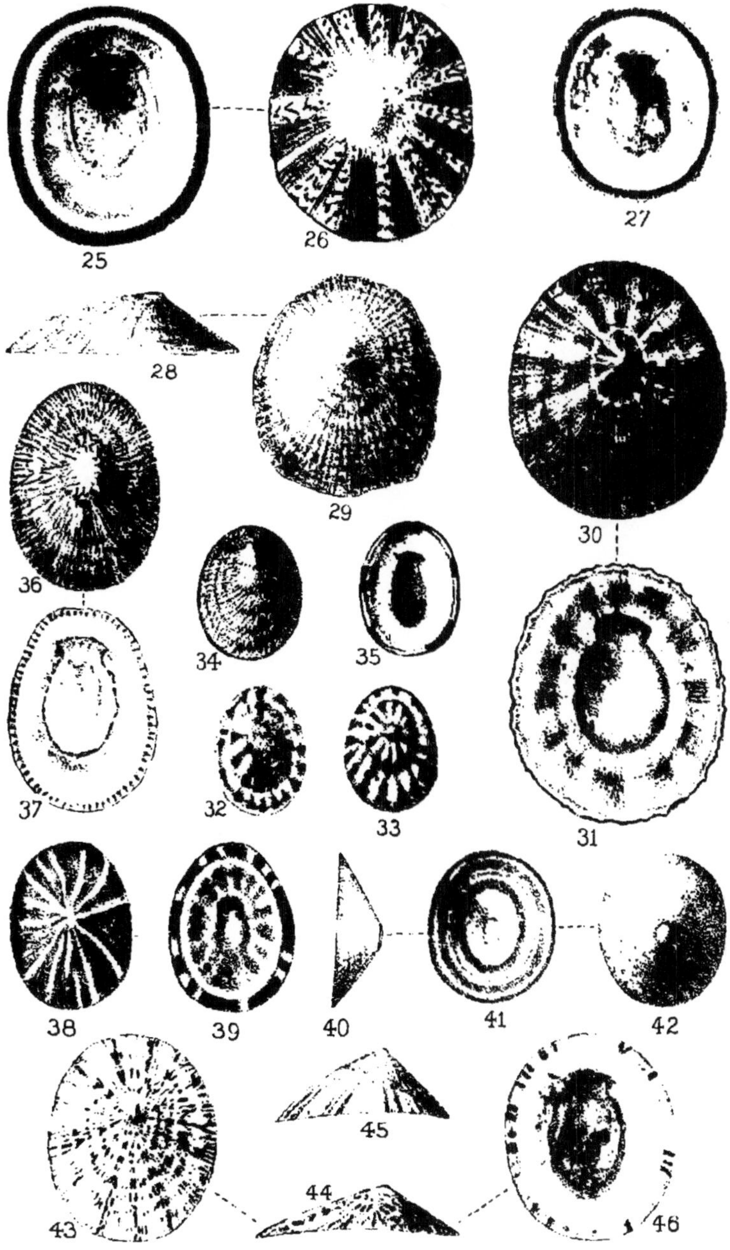
25
26
27
28
29
30
36
34
35
37
32
33
31
38
39
40
41
42
43
45
44
46

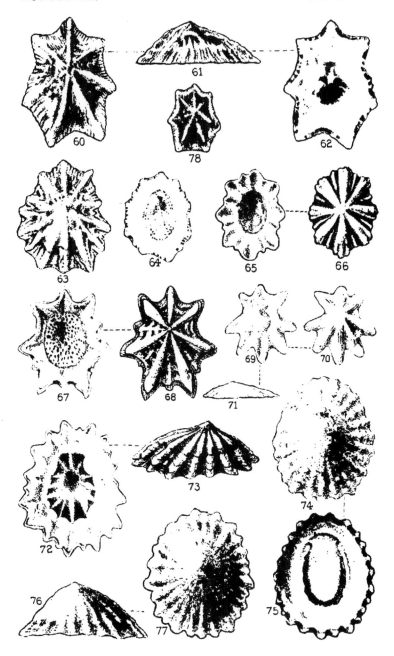

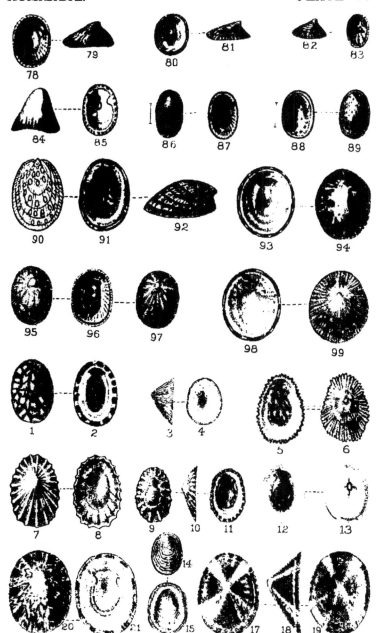

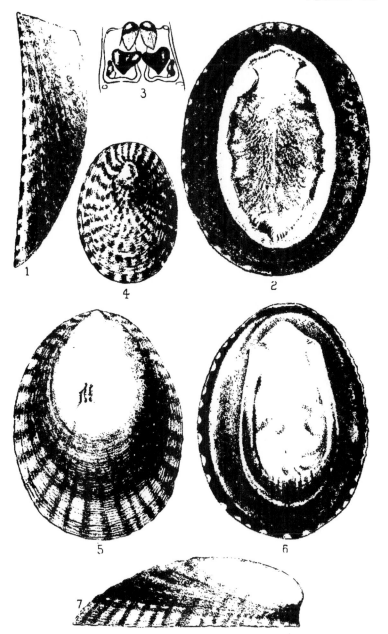

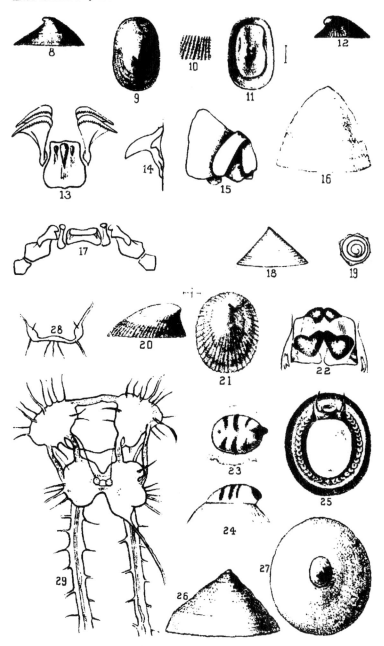

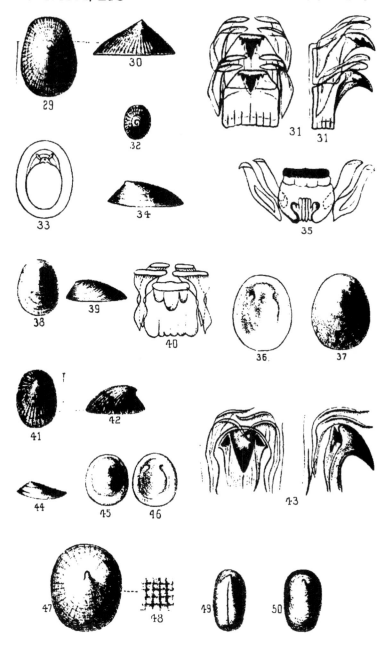

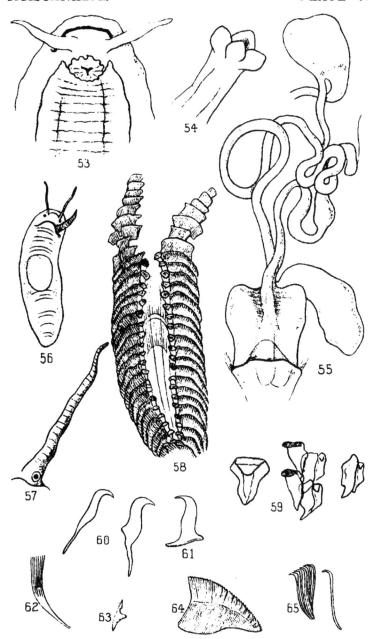

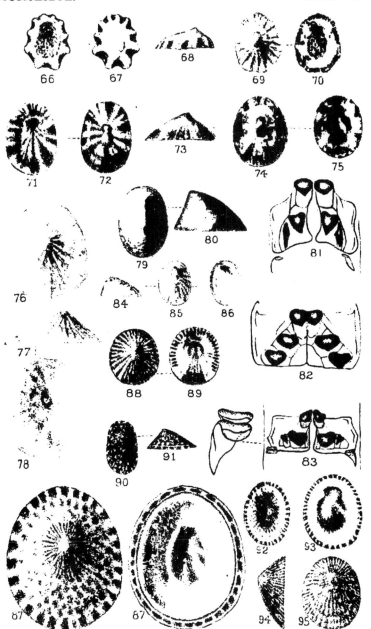

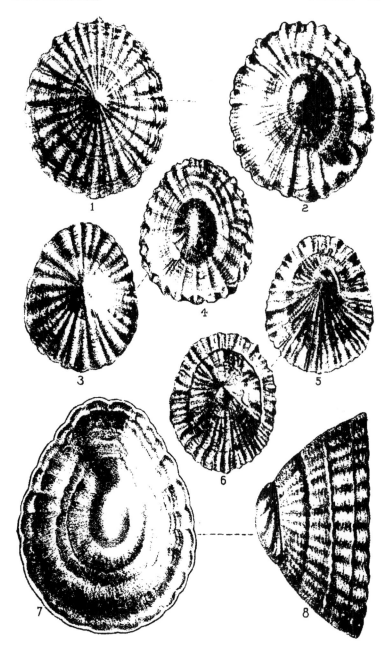

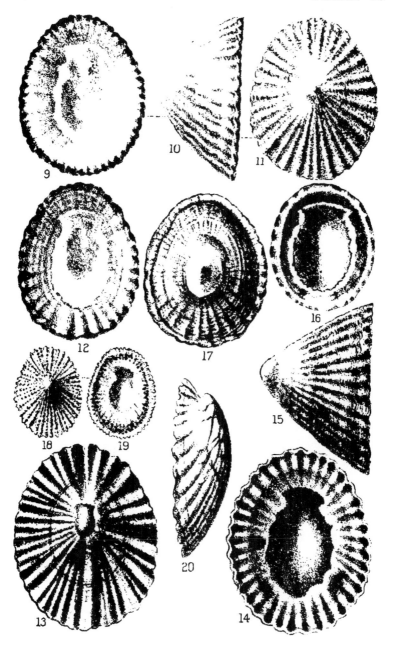

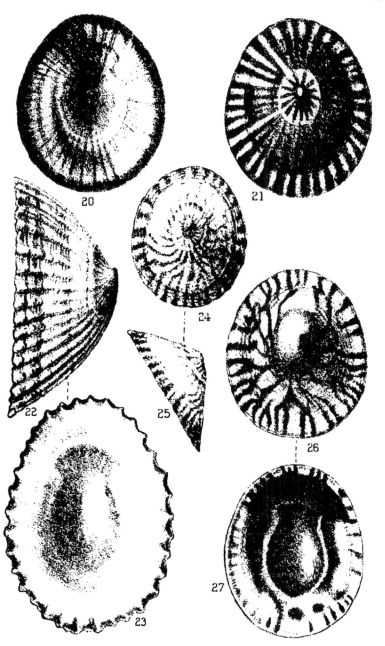

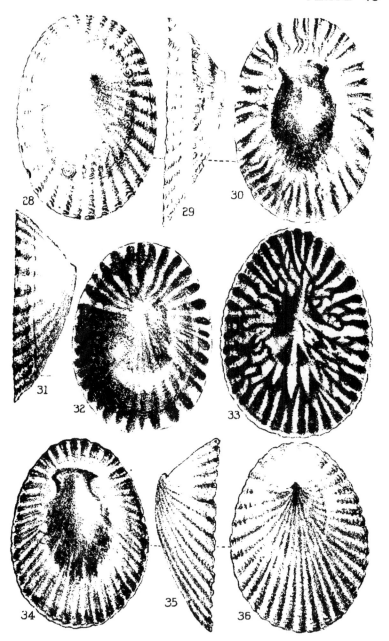

PLATE 48

PATELLIDÆ.

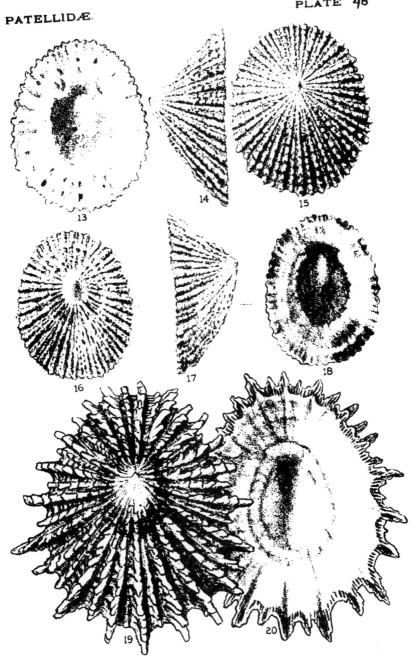

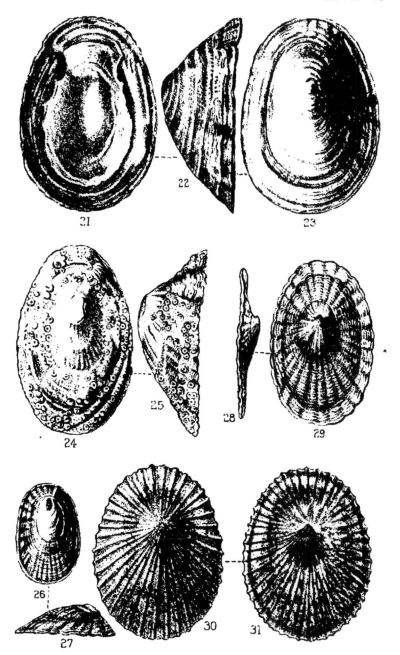

21   22   23

24   25   28   29

26   27   30   31

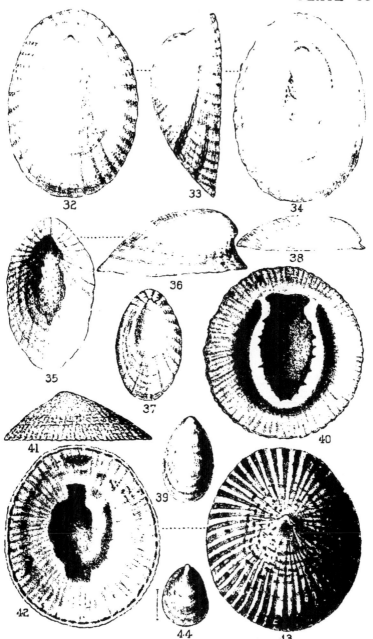

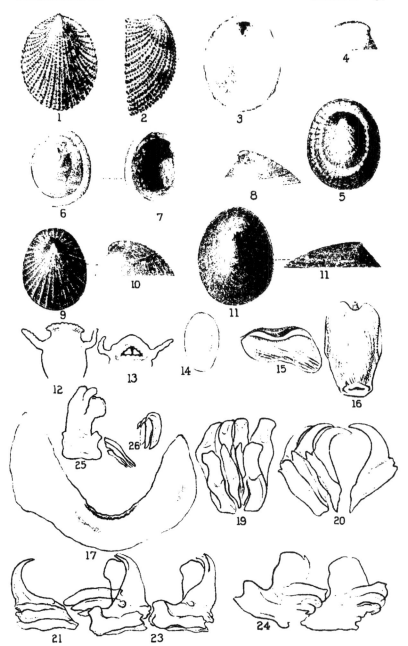

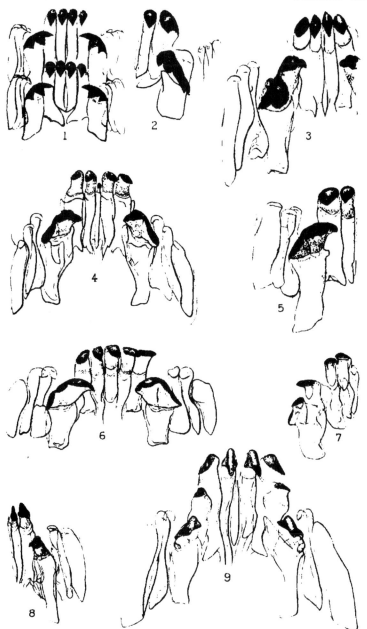

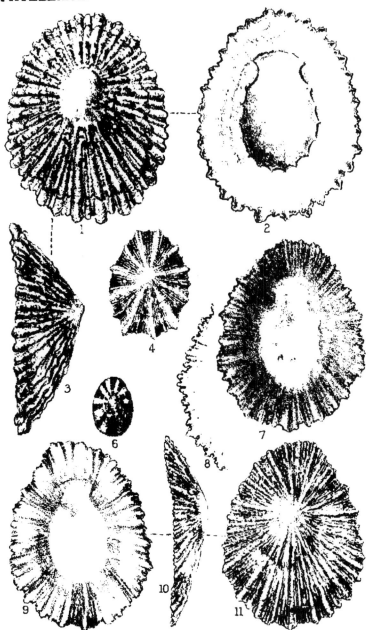

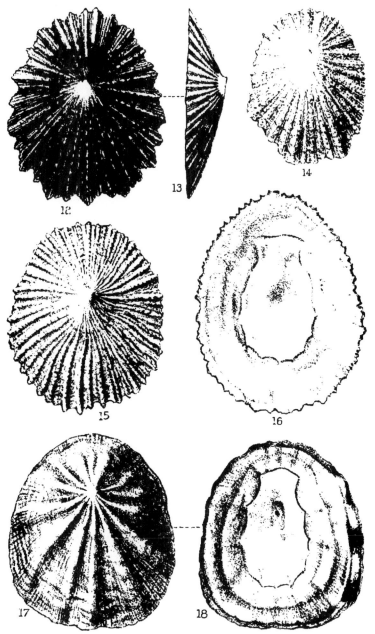

12 13 14

15 16

17 18

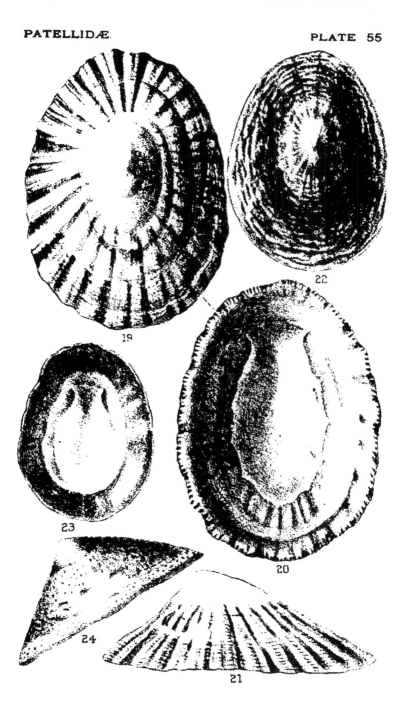

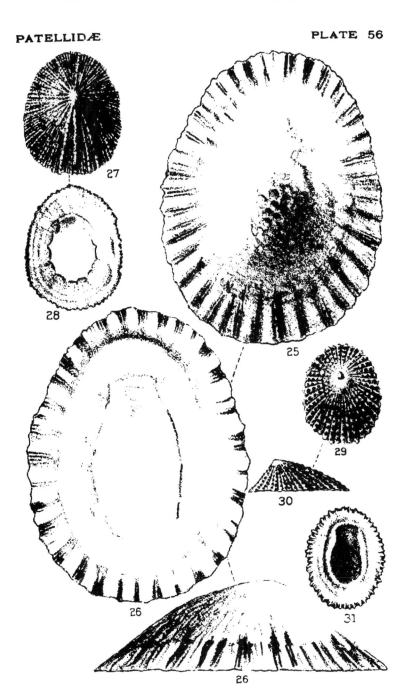

27

28

25

29

30

26

31

26

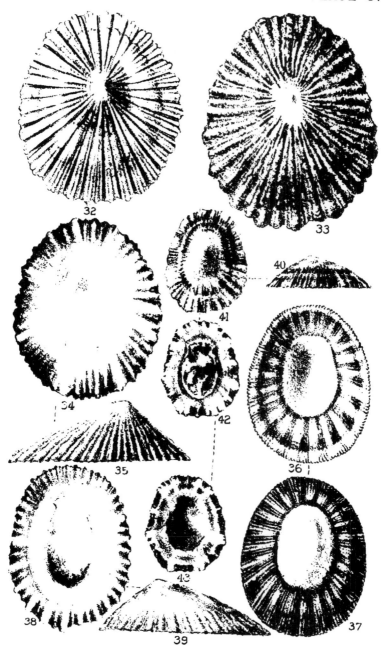

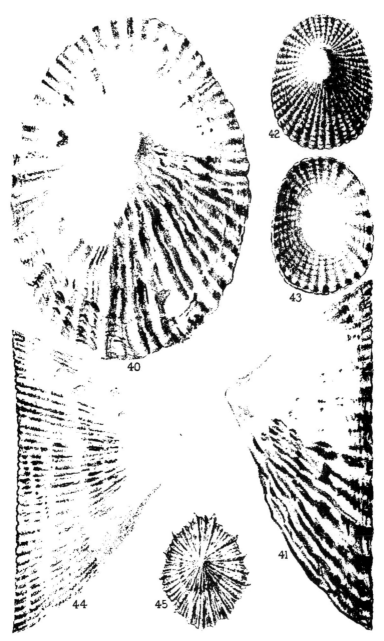

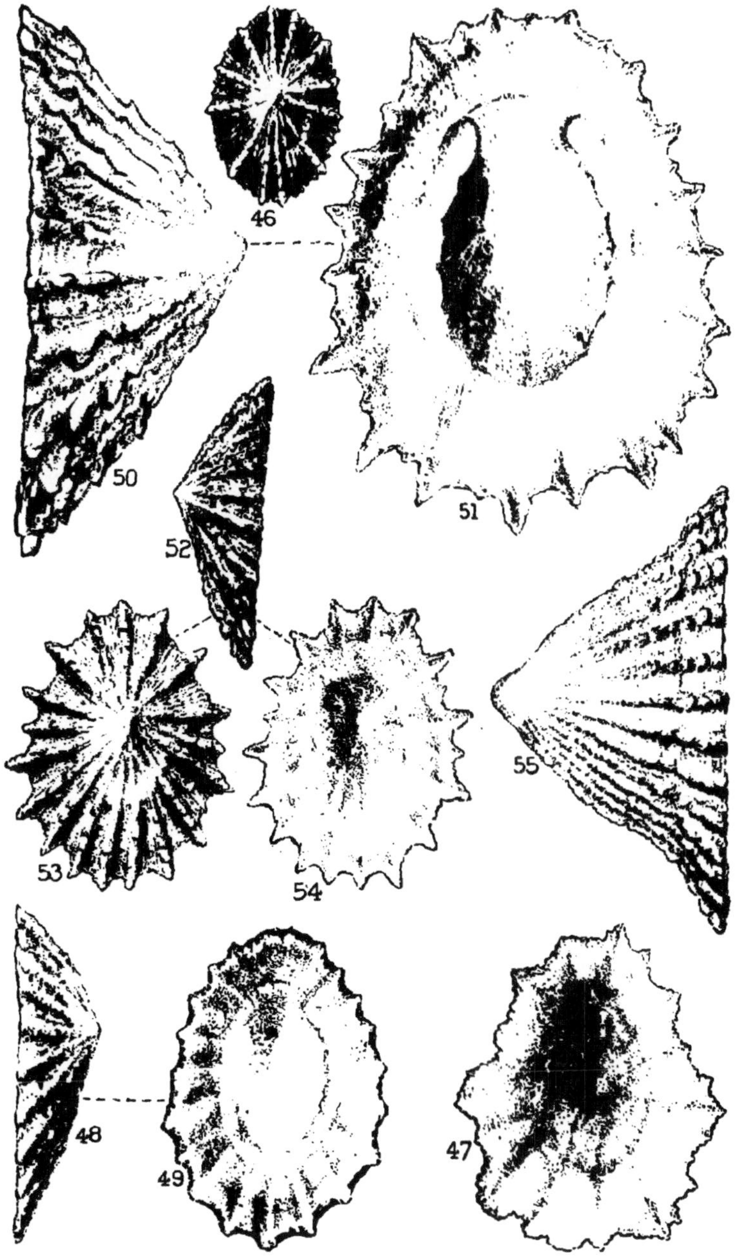
46
50
51
52
53
54
55
48
49
47

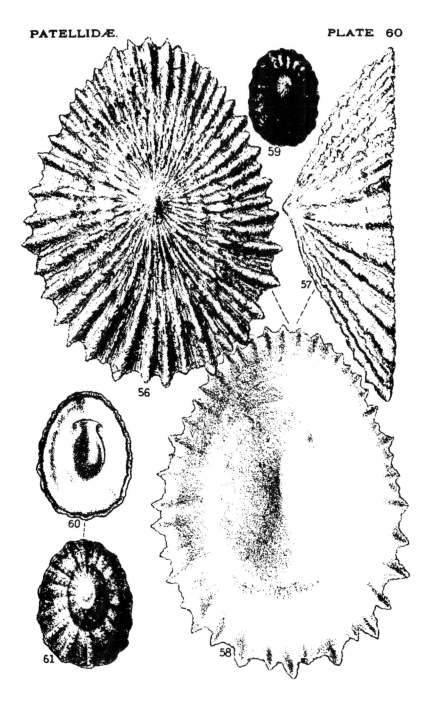

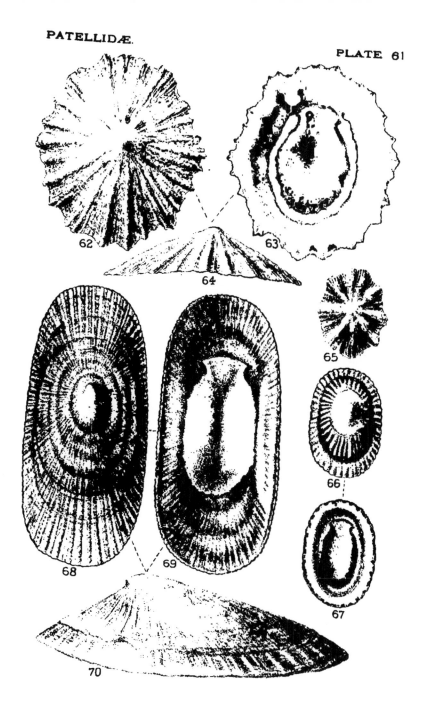

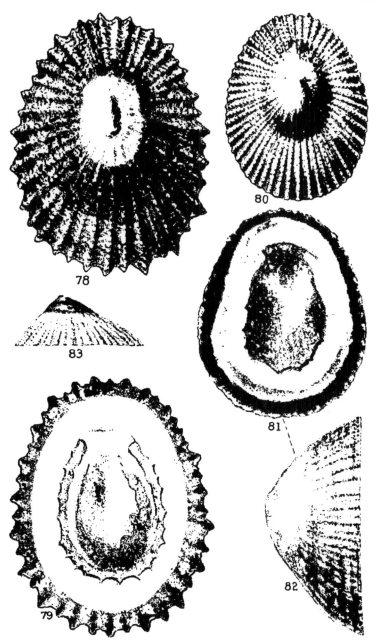

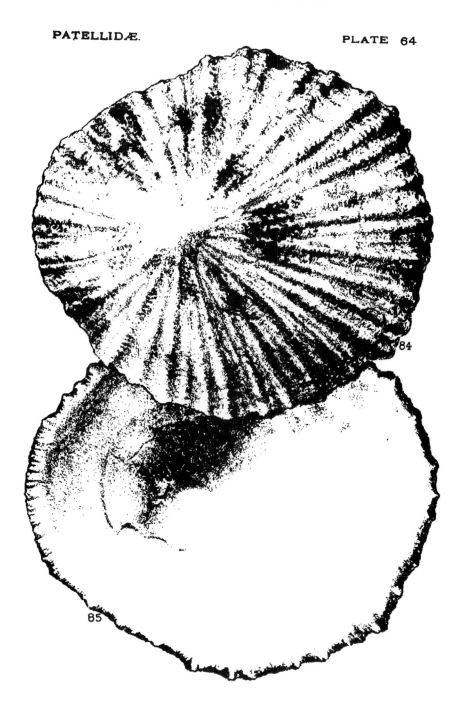

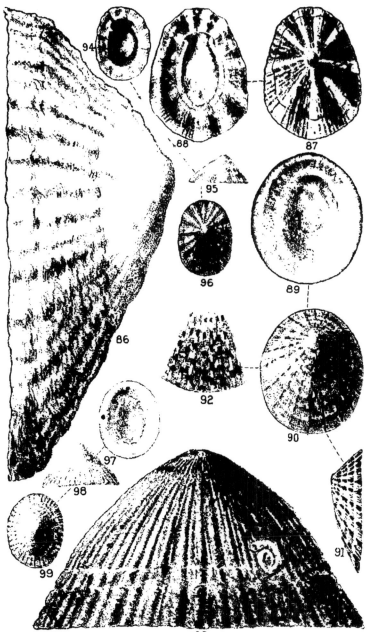

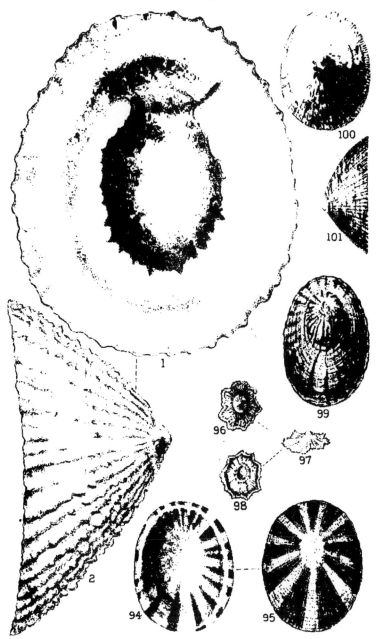

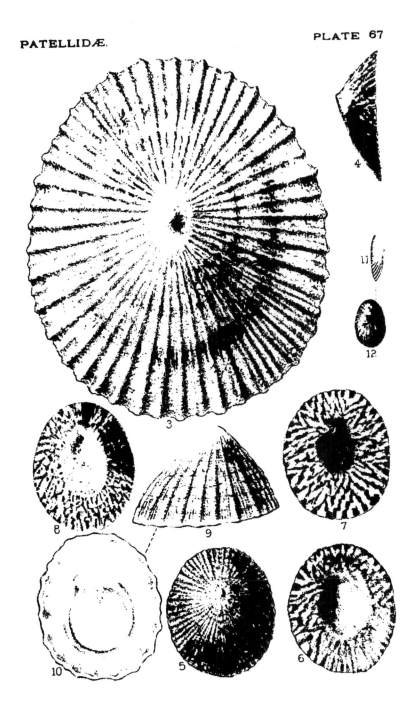

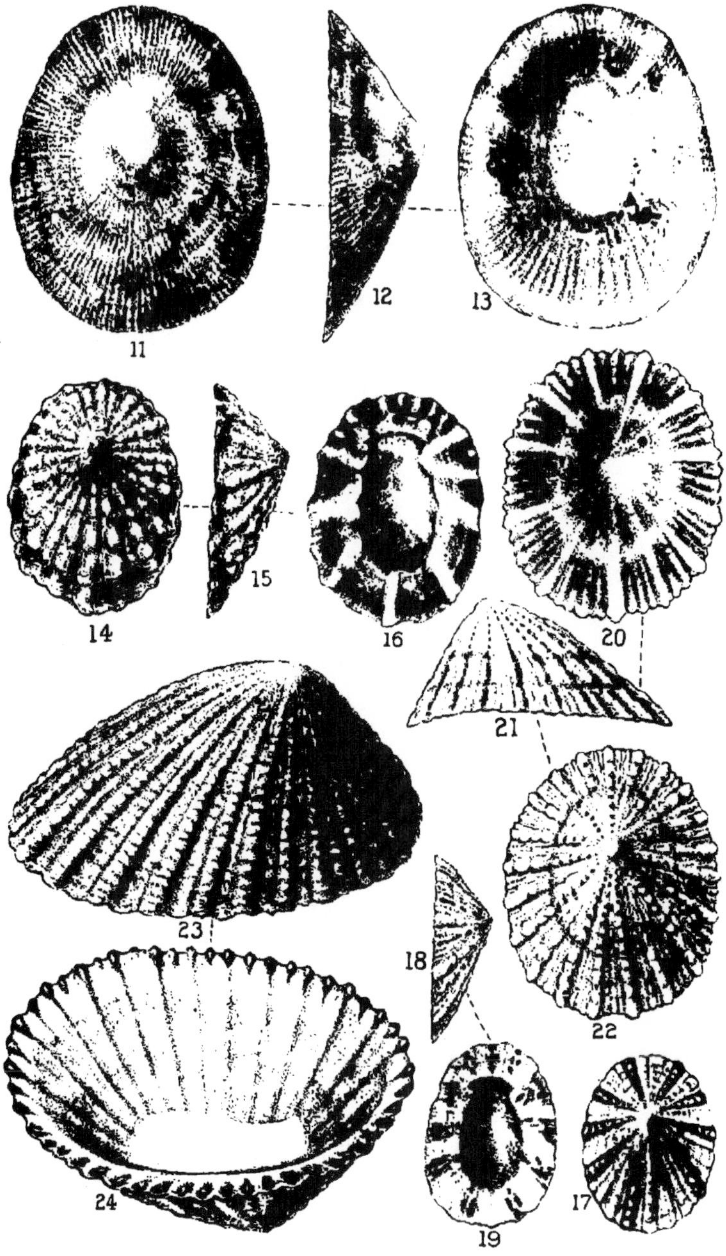
11
12
13
14
15
16
20
21
23
18
22
24
19
17

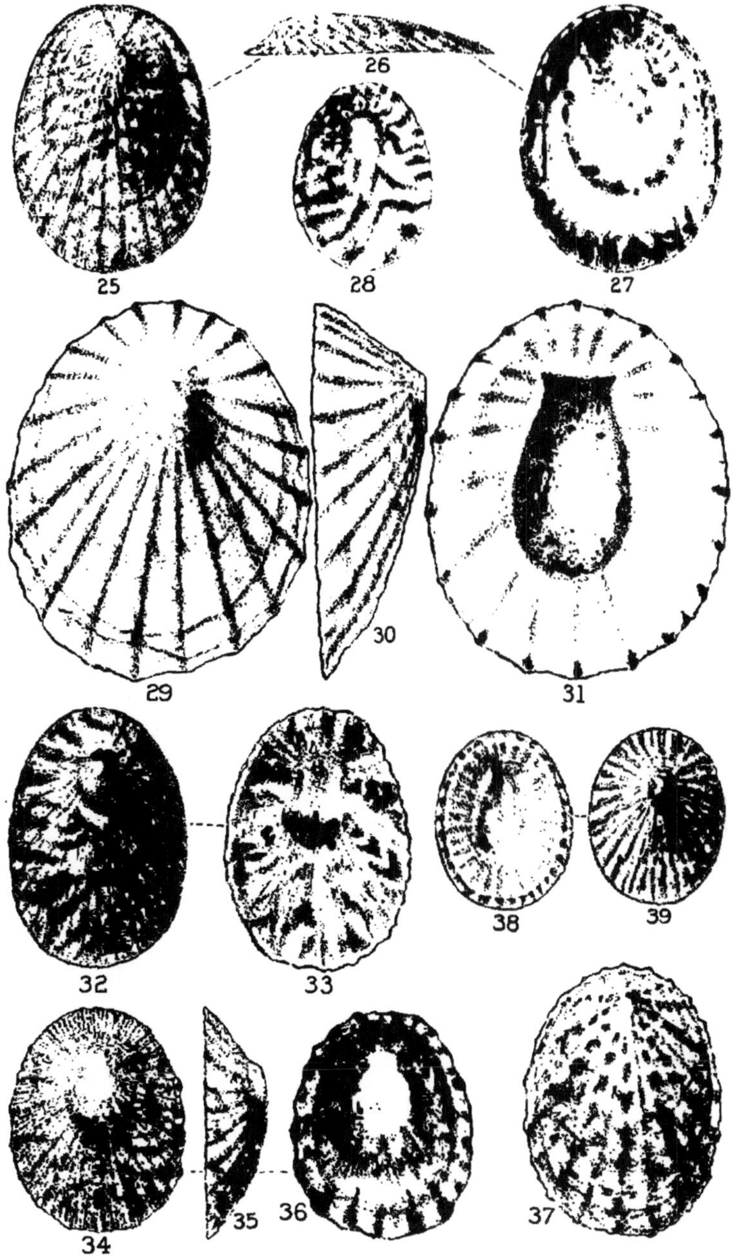
26
25
28
27
30
29
31
32
33
38
39
35
36
37
34

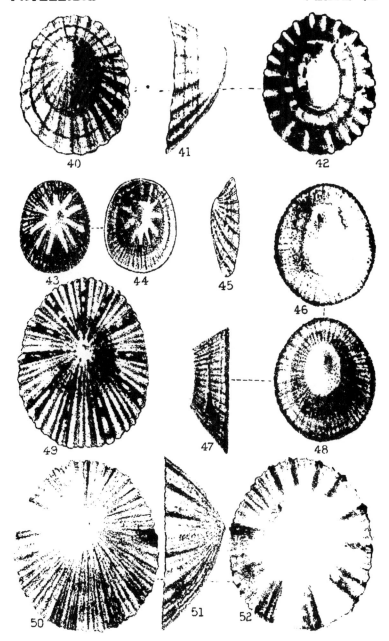

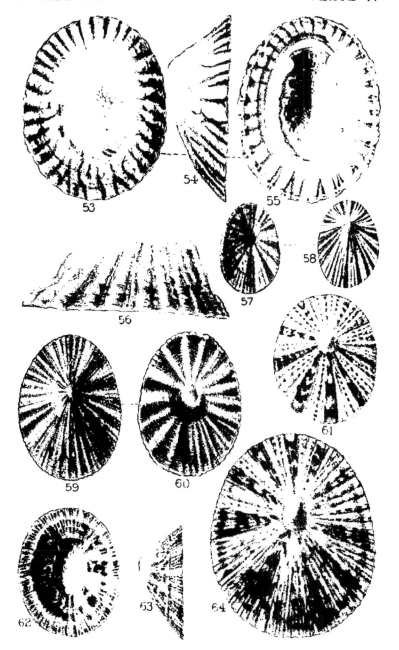

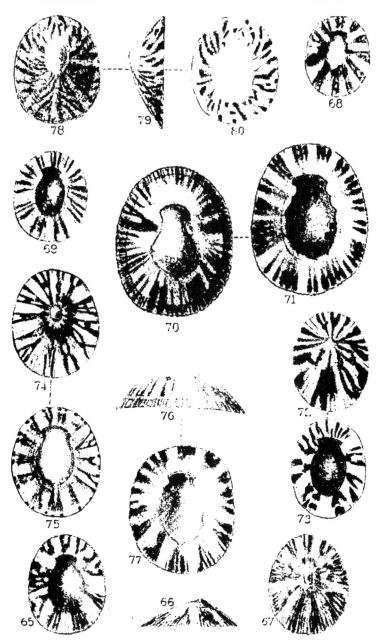

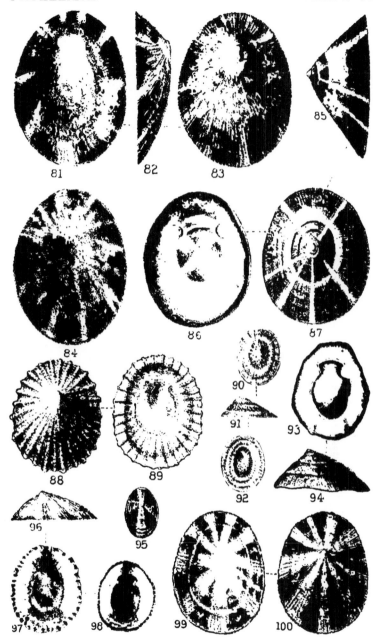

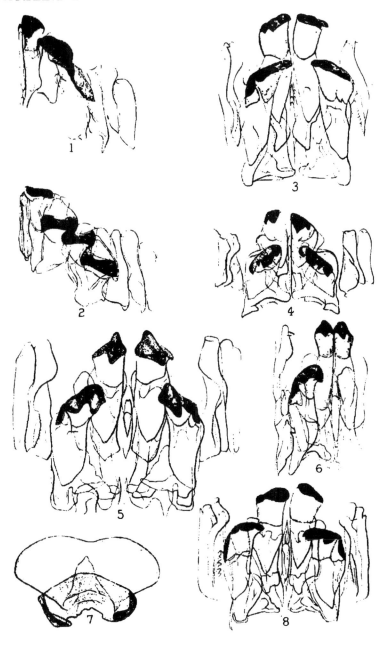